T0143302

Applications of Machine Learning in Big-Data Analytics and Cloud Computing

RIVER PUBLISHERS SERIES IN INFORMATION SCIENCE AND TECHNOLOGY

Indexing: All books published in this series are submitted to the Web of Science Book Citation Index (BkCI), to SCOPUS, to CrossRef and to Google Scholar for evaluation and indexing.

The "River Publishers Series in Information Science and Technology" covers research which ushers the 21st Century into an Internet and multimedia era. Multimedia means the theory and application of filtering, coding, estimating, analyzing, detecting and recognizing, synthesizing, classifying, recording, and reproducing signals by digital and/or analog devices or techniques, while the scope of "signal" includes audio, video, speech, image, musical, multimedia, data/content, geophysical, sonar/radar, bio/medical, sensation, etc. Networking suggests transportation of such multimedia contents among nodes in communication and/or computer networks, to facilitate the ultimate Internet.

Theory, technologies, protocols and standards, applications/services, practice and implementation of wired/wireless networking are all within the scope of this series. Based on network and communication science, we further extend the scope for 21st Century life through the knowledge in robotics, machine learning, embedded systems, cognitive science, pattern recognition, quantum/biological/molecular computation and information processing, biology, ecology, social science and economics, user behaviors and interface, and applications to health and society advance.

Books published in the series include research monographs, edited volumes, handbooks and textbooks. The books provide professionals, researchers, educators, and advanced students in the field with an invaluable insight into the latest research and developments.

Topics covered in the series include, but are by no means restricted to the following:

- Communication/Computer Networking Technologies and Applications
- Queuing Theory
- Optimization
- Operation Research
- Stochastic Processes
- Information Theory
- Multimedia/Speech/Video Processing
- Computation and Information Processing
- Machine Intelligence
- Cognitive Science and Brian Science
- Embedded Systems
- Computer Architectures
- Reconfigurable Computing
- Cyber Security

For a list of other books in this series, visit www.riverpublishers.com

Applications of Machine Learning in Big-Data Analytics and Cloud Computing

Editors

Subhendu Kumar Pani

Principal, Krupajal Computer Academy, India

Somanath Tripathy

Indian Institute of Technology, India

George Jandieri

Georgian Technical University, Russia

Sumit Kundu

National Institute of Technology Durgapur, India

Talal Ashraf Butt

The American University in the Emirates, UAE

River Publishers

Published, sold and distributed by:
River Publishers
Alsbjergvej 10
9260 Gistrup
Denmark

www.riverpublishers.com

ISBN: 978-87-7022-182-5 (Hardback)
978-87-7022-181-8 (Ebook)

Contents

Preface

This book comprehensively covers the topic of techniques and applications of machine learning in big-data analytics and cloud computing is an emerging field of research at the intersection of information science and computer science. Cloud computing and big data technologies have moved out from the realm of hype to one of the heart descriptors of the new digital age. The digital data have propagated nine times in volume in just five years and by 2030, scientists are suggesting that its volume will increase at a rapid rate and will almost reach 65 trillion gigabytes. In this paradigm shift, a investigate portrays that, 90% of all the data in the world was generated in this past 2 years and it is obvious that we are breathing a data cascade era. This explosion of data igniting trend embarks ample opportunities and considerable transformation in various areas such as healthcare, enterprises, industrial manufacturing, and Transportation. Cloud and Big data endows a new technique of probability to collect, manage and analyze the vast quantities of data, which indeed offers an understanding of context towards varied applications. In cloud and big data analytics, Swarm Intelligence and deep learning is a set of machine-learning techniques is still budding but shows enormous potential for solving business problems. Deep learning facilitates computers to be familiar with items of concern in large quantities of unstructured and binary data and to deduce relationships without requiring specific models or programming instructions. Furthermore, deep learning has an integral part in the advancement in Computer Vision, speech recognition, Image Processing and Natural Language Processing in the last decade. Using deep learning techniques few extensive standing issues were sorted out in speech recognition.

The book will play a vital role in real time environment to a great extent. All the researchers and practitioners will be highly benefited those are working in field of cloud computing, machine learning and big data analytics. This book would be a good collection of state-of-the-art approaches for machine learning based big data and cloud related applications. It will be very beneficial for the new researchers and practitioners working in the field to quickly know the best performing methods. They would be able to compare different approaches and can carry forward their research in the most important area of research which has direct impact on solving the real world problems This book would be very useful because there is no book in the market which provides a good collection of state-of-the-art methods of machine learning based big data and cloud related applications.

Machine learning based big data and cloud related applications are recently emerged and very un-matured field of research in computer science, Machine learning, biomedical and healthcare. This book covers the latest advances and developments in the field of data mining, machine learning, and cloud computing.

Organization of the Book

The 14 chapters of this book present scientific concepts, frameworks and ideas on data analytics and cloud computing using different machine learning techniques. The Editorial Advisory Board and expert reviewers have ensured the high caliber of the chapters through careful refereeing of the submitted chapters. For the purpose of coherence, we have organized the chapters with respect to similarity of topics addressed, ranging from machine learning issues pertaining to pattern analysis, wireless multimedia sensor network, Spam Detection and Health Informatics.

Chapter 1, "Pattern Analysis of COVID-19 Death and Recovery Cases Data of Countries Using Greedy Biclustering Algorithm" S. Dhamodharavadhani and R. Rathipriya present the biclustering approach to analyze the COVID-19 data.This is a new application of biclustering approach to analyze COVID-19 data. The proposed approach can alert and help the government authorities and healthcare professionals to know what to anticipate and which measures to implement to decelerate the COVID-19 spread.

Chapter 2, "Artificial Fish Swarm Optimization Algorithm with Hill Climbing Based Clustering Technique for Throughput Maximization in Wireless Multimedia Sensor Network" N. Krishnaraj, T. Jayasankar, N. V. Kousik, A. Daniel focuses new artificial fish swarm optimization algorithm (AFSA) with hill climbing (HC) based clustering technique for throughput maximization in WMSN, called AFSA-HC. The proposed AFSA-HC algorithm involves four main processes to maximize throughput in the network, namely node initialization, AFSA-HC based node clustering, deflate algorithm based data aggregation, and hybrid data transmission.

A comprehensive experimental analysis is carried out to verify the effective performance of the presented AFSA-HC technique. The simulation outcome inferred that the AFSA-HC technique has attained maximum results in terms of different measures, namely energy consumption, throughput, network lifetime, network stability, and packet loss.

In Chapter 3, "Analysis of Machine Learning Techniques for Spam Detection" Supriya Raheja, Shreya Kasturia discusses the spam detection and different machine learning models to detect these spam messages. The present work also discusses the testing and evaluation of spam messages. Spam messages have always been very dangerous for computers and networks. They have a very bad effect on the computer security. With the emergence of social media platforms, many people are dependent on emails to communicate, and, with this, there is always a need to detect and prevent spam mails before it enters a user's inbox. The paper also presents the analyses of different machine learning techniques to detect spam messages. Finally, the paper describes the algorithm which is best to detect spam messages.

In Chapter 4, "Smart Sensor Based Prognostication of Cardiac Disease Prediction Using Machine Learning Techniques" Vijayaganth V., Naveen kumar M. present machine learning (ML) algorithms to implement a solution for earlier prediction of cardiovascular disease. The work comprises various sensors to get the user's medical parameter like resting blood pressure (RBS), serum cholesterol (chol), fasting blood sugar (FBS), max heart rate, etc., and the data are processed with various methods and compared the accuracy of the disease prediction.

In Chapter 5, "Assimilate Machine Learning Algorithms in Big Data Analytics: Review" Sona D. Solanki, Asha D. Solanki, Samarjeet Borah focus on the major complexities of integrating ML algorithms to the Big Data framework and demonstrates an innovative strategy to enable massive data analysis of such parameters' productivity and convenience. This chapter summarizes a literature overview of the various ML techniques. Moreover, a review is performed and discussed in this document on popularly utilized ML strategies for significant dataset insights. Implementing ML models a new methodology is utilized for scientific and health disciplines, data collection, categorization and reconstruction on large datasets that offers a solid platform and inspiration for further work in the area of data management ML.

In Chapter 6, "Resource Allocation Methodologies in Cloud Computing: A Review and Analysis" Pandaba Pradhan, Prafulla Ku. Behera, B. N. B. Ray present the review on current competitive resource allocation methods in cloud computing. Cloud computing has emerged as a new technology that has got huge potential in enterprises and markets today. Clouds can make it possible to access applications and associated data from anywhere. Companies are able to rent resources from cloud for storage and other computational purposes so that their infrastructure cost can be reduced significantly. Cloud technology is an on-demand infrastructure platform that provides pay-per-use services to cloud users. In order to efficiently allocate resources and to meet the needs of users, an effective and flexible allocation of resources mechanism is required. According to the growing requirements of the customers, the resource allocation process has been more complex and complicated. One of the main priorities of research scholars is on how to design effective solutions for this problem.

In Chapter 7, "Role of Machine Learning in Big Data" Nilanjana Das*, Santwana Sagnika and Bhabani Shankar Prasad Mishra discuss about implementation of machine learning techniques in big data. Machine learning is promising when large-scale data processing is considered as it has the capability of learning from the previous data and incorporating those findings on new incoming data. Although its applications in big data are limitless, it faces a vast number of challenges.

In Chapter 8, "Healthcare System for COVID-19: Challenges and Developments " Arun Kumar, Dr. Sharad Sharma provides some solution developments with challenges Of IoT in healthcare for COVID-19.The Internet of Healthcare Things is, in a general sense, changing how individuals speak with one another and with their general surroundings. Internet of Things (IoT) is the propelled organized framework of the network, transportation, and innovation. IOT brilliant gadgets can execute the offices of remote wellbeing checking and crisis notice framework. IoT has obvious utilization of shrewd medicinal services framework

In Chapter 9, "An Integrated Approach of Blockchain & Big Data in Health Care Sector" Nitin Tyagi, Bharat Bhushan, Siddharth Gautam, Nikhil Sharma, Santosh Kumar discuss an overview of the blockchain that includes the Health Care Big Data, as well as blockchain solutions for big data on healthcare. In this chapter, a holistic model of the healthcare system, specifically for developing a smart healthcare.Blockchain has many inherent features like decentralized storage, authentication, distributed ledger, immutability and safety, which are more practical in the fields of healthcare.

In Chapter 10, "Cloud Resource Management for Network Cameras" Hemanta Kumar Bhuyan, Subhendu Kumar Pani" have an important discussion about network cameras flow and adequate visual data from several locations such as traffic, check gate/tollgate, natural scenes, etc.These data from high resolution network cameras can be used in several applications that help to find the needed significant resources. The cloud vendors provide the required resources as per the hourly cost which is not always so affordable for several applications using cameras and instances. This problem can be solved using bin packing and heuristic algorithm.

In Chapter 11, "Software-Defined Networking for Healthcare Internet of Things" Rinki Sharma discusses the advantages, challenges, and possible solutions of incorporating software definednetworks (SDN) in healthcare Internet of Things (H-IoT). An architecture for SDN-based H-IoT system is presented and discussed. The chapter also discusses open research challenges, possible solutions, and research opportunities of integrating SDN with H-IoT..

In Chapter 12, "Cloud Computing in the Public Sector: A Study". Amita Verma, Anukampa present in detail what exactly cloud computing is, its evolution, and how this brilliant technology can be used in the various agencies of the public sector along with the advantages of using it, which include easy sharing between different agencies, cost cutting, etc. and the challenges that come with it.

In Chapter 13, "Big Data Analytics: An overview" Dipalika Das, Maya Nayak discover different analytic tools and techniques that would be beneficial while dealing with big data in various fields. Today we are in an information era where large amount of data are available for decision making. Those datasets which are huge in volume, fast in velocity, and large in variety are known as big data. Existing tools and technologies are not sufficient enough to handle such a big data and, hence, would face difficulty.

In Chapter 14 "Video Usefulness Detection in Big Surveillance Systems" Hemanta Kumar Bhuyan, Subhendu Kumar Pani present the different failure detection approaches are explained to determine the malfunction during video data transmission. Finally, few technical methods elaborated the predicted video items on true or false basis evaluation VU model.

The book is a collection of the fourteen chapters by eminent professors, researchers, and industry people from different countries. The chapters were initially peer reviewed by the editorial board members, reviewers, and industry people who themselves span many countries. The chapters are arranged so that all the chapters have the basic introductory topics and the advances as well as future research directions, which enable budding researchers and engineers to pursue their work in this area.

Machine learning in big-data analytics and cloud computing are so diversified that it cannot be covered in single book. However, with the encouraging research contributed by the researchers in this book, we (contributors), Editorial members, and reviewers tried to sum up the latest research domains, development in the data analytics field, and applicable areas. First and foremost we express heartfelt appreciation to all authors. We thank them all in considering and trusting this Edited Book as the platform for publishing their valuable work. We also thank all authors for their kind co-operation extended during the various stages of processing of the manuscript. This edited book will serve as a motivating factor for those researchers who have spent years working as machine learning experts, data analysts, statisticians, and budding researchers.

List of Contributors

Anukampa, *Research scholar, Kurukshetra University, Haryana, India; E-mail: anukampa001@gmail.com*

Prafulla Ku. Behera, *Reader & H.O.D, Department of Computer Science & Applications, Utkal University, BBSR, Odisha, India; E-mail: p_behera@hotmail.com*

Bharat Bhushan, *School of Engineering and Technology (SET), Sharda University, India; E-mail: bharat_bhushan1989@yahoo.com*

Samarjeet Borah, *Sikkim Manipal Institute of Technology, Sikkim Manipal University, Sikkim, India; E-mail: samarjeetborah@gmail.com*

Talal Ashraf Butt, *The American University in the Emirates (AUE), UAE.*

A. Daniel, *Assistant Professor, School of Computing Science and Engineering, Galgotias University, Greater Noida, Uttar Pradesh, India; E-mail: danielarockiam@gmail.com*

Dipalika Das, *Department of MCA, Trident Academy of Creative Technology, BBSR, Odisha, India; E-mail: dipalika.das@gmail.com*

Nilanjana Das, *School of Computer Engineering, Kalinga Institute of Industrial Technology Deemed to be University, Bhubaneswar, Odisha – 751024, India; E-mail: nilanjanadas010@gmail.com*

S. Dhamodharavadhani, *Department of Computer Science, Periyar University, Salem, India; E-mail: vadhanimca2011@gmail.com*

Siddharth Gautam, *Raj Kumar Goel Institute of Technology, Ghaziabad, U.P.; E-mail: siddharthinfo92@gmail.com*

George Jandieri, *Chief scientist at Georgian Technical University, Russia (USSR).*

T. Jayasankar, *Assistant Professor, Department of ECE, University College of Engineering, BIT Campus, Anna University, Tiruchirappalli-620024, Tamilnadu, India; E-mail: jayasankar27681@gmail.com*

Arun Kumar, *Panipat Institute of Engineering and Technology, Samalkha; E-mail: ranaarun1.ece@piet.co.in*

Hemanta Kumar Bhuyan, *Department of Information Technology, Vignan's Foundation for Science, Technology & Research (Deemed to be University), Guntur, Andhra Pradesh, India, PIN-522213; E-mail: hmb.bhuyan@gmail.com*

N. Krishnaraj, *Associate Professor, School of Computing, SRM Institute of Science and Technology, Kattankulathur 603203, Tamilnadu, India; E-mail: drnkrishanraj@gmail.com*

N. V. Kousik, *Associate Professor, School of Computing Science and Engineering, Galgotias University, Greater Noida, Uttar Pradesh, India; E-mail: nvkousik@gmail.com*

Santosh Kumar, *ITER-SOA, Bhubaneswar, Odisha; E-mail: santoshkumar@soa.ac.in*

Shreya Kasturia, *Amity University, Noida*

Sumit Kundu, *Professor of ECE at NIT Durgapur, India*

M. Naveenkumar, *Assistant Professor, Department of Computer Science and Engineering, KPR Institute of Engineering and Technology, Coimbatore, India; E-mail: naveenkumar2may94@gmail.com*

Maya Nayak, *Department of CSE, Orissa Engineering College, BBSR, Odisha, India; E-mail: mayanayak3299@yahoo.com*

Pandaba Pradhan, *Asstistant Professor & H.O.D, Department of Computer Science, BJB (Auto) College, BBSR, Odisha, India; E-mail: ppradhan11@gmail.com*

Subhendu Kumar Pani, *Principal, Krupajal Computer Academy, India; E-mail: skpani.india@gmail.com*

B. N. B. Ray, *Reader, Department of Computer Science & Applications, Utkal University, BBSR, Odisha, India; E-mail: bnbray@yahoo.com*

R. Rathipriya, *Department of Computer Science, Periyar University, Salem, India; E-mail: rathipriyar@gmail.com*

Supriya Raheja, *Amity University, Noida; E-mail: supriya.raheja@gmail.com*

Bhabani Shankar Prasad Mishra, *School of Computer Engineering, Kalinga Institute of Industrial Technology Deemed to be University, Bhubaneswar, Odisha – 751024, India; E-mail: mishra.bsp@gmail.com*

Nikhil Sharma, *Ambedkar Institute of Advanced Communication Technologies & Research, Delhi; E-mail: nikhilsharma1694@gmail.com*

Rinki Sharma, *Ramaiah University of Applied Sciences; E-mail: rinki.cs.et@msruas.ac.in*

Santwana Sagnika, *School of Computer Engineering, Kalinga Institute of Industrial Technology Deemed to be University, Bhubaneswar, Odisha – 751024, India; E-mail: santwana.sagnika@gmail.com*

Sharad Sharma, *Maharishi Markandeshwar (Deemed to be University), Mullana; E-mail: sharadpr123@rediffmail.com*

Asha D. Solanki, *Mahila Jagruti News, Vadodara, India; E-mail: solankiasha2710@gmail.com*

Sona D. Solanki, *Department of Electronics and Communication Engineering, Babaria Institute of Technology, Vadodara, India; E-mail: solankisona28@gmail.com*

Nitin Tyagi, *HMR institute of Technology & Management, Delhi; E-mail: nitintyagi5631@gmail.com*

Somanath Tripathy, *Indian Institute of Technology, India*

Amita Verma, *Associate Professor, University Institute of Legal Studies, Panjab University, Chandigarh, India; E-mail: amitaverma21@gmail.com*

V. Vijayaganth, *Assistant Professor, Department of Computer Science and Engineering, KPR Institute of Engineering and Technology, Coimbatore, India; E-mail: kv.vijayaganth@gmail.com*

List of Figures

List of Tables

List of Abbreviations

3D	Three Dimensional
5G	Fifth Generation
AAA	Authentication, Authorization and Accounting
AES	Advanced Encryption Standard
API	Application Program Interface
CAPEX	Capital Expenditure
CCTV	closed-circuit television
DAS	Direct Attached Storage
DBP	Dynamic Bin Packing
ECDH	Elliptic-Curve Diffie-Hellman
ENT	Ear, Nose, Throat
FN	False Negative
FP	False Positive
GPS	Global Positioning System
GSM	Global System for Mobile communication
H-IoT	Healthcare Internet of Things
ID	Identifier
IEEE	Institute of Electrical and Electronics Engineers
IoE	Internet Of Everything
IoT	Internet of Things
IoT	Internet Of Things
IP	Internet Protocol
IPv6	Internet Protocol version 6
IrDA	Infrared Data Association
LTE	Long Term Evolution
M2M	Machine - to - Machine
MTTD	Mean Time To Detection
NAS	Network Attached Storage
NetCam	Network camera ()
NetCam	Network Camera
NFV	Network Function Virtualization

NVR	network video recorder
OPEX	Operational Expenditure
PB	Petabytes
QoS	quality of service
QoS	Quality of Service
QoS	Quality Of Service
QoS	Quality Of User Experience
RFID	Radio Frequency Identification
RIC	Resource Information Centre
RPA	Resource Provisioning Agent
SAN	Storage Area Network
SDN	Software Defined Networking
SLA	service level agreement
SSID	Service Set Identifier
TN	True Negative
TP	True Positive
UWB	Ultra - Wideband
VM	virtual machines
VSS	Several Video Surveillance Systems
VU	Video Usefulness
WAN	Wide Area Network
WEP	Wired Equivalent Privacy
Wi-Fi	Wireless Fidelity
WPA	Wi-Fi Protected Access
WRM	Workload Resource Manager
WSN	Wireless Sensor Networks

1

Pattern Analysis of COVID-19 Death and Recovery Cases Data of Countries Using Greedy Biclustering Algorithm

S. Dhamodharavadhani and R. Rathipriya

Department of Computer Science, Periyar University, Salem, India
Email: vadhanimca2011@gmail.com; rathipriyar@gmail.com

Abstract

COVID-19 data were analyzed using the biclustering approach to gain insights such as which groups of countries have similar epidemic trajectory patterns over the subset of COVID-19 pandemic outburst days (called bicluster). Countries within these groups (biclusters) are all in the same phase but with a slightly different trajectory. An approach based on the Greedy Two Way KMeans biclustering algorithm (also called Greedy Biclustering) is proposed to analyze COVID-19 epidemiological data, which identifies subsets of countries that represent a similar epidemic trajectory pattern over a specific period of time. To the best of authors' knowledge, this is a new application of biclustering approach to analyze COVID-19 data. Results confirm that the proposed approach can alert and help the government authorities and healthcare professionals to know what to anticipate and which measures to implement to decelerate the COVID-19 spread.

Keywords: COVID-19 data, biclustering, Greedy Two Way KMeans biclustering, COVID-19 pattern, algorithm, death data, recovery data.

1.1 Introduction

Since 2000, biclustering methods are a very popular data mining technique for pattern analysis in the field of bio-informatics. "These biclustering methods are used to group the rows and columns of a data matrix simultaneously" [1, 2]. This work investigates the applicability of biclustering methods to daily cumulative COVID-19 death and recovery cases data of 50 countries around the world. While clustering is applied to group COVID-19 cases, it has the following limitations:

- It clusters either countries or days (pandemic outburst days) of COVID data that is only one at a time (i.e., either row-wise or column-wise).
- It is unable to identify the subset of countries that have similar epidemic spread trends over a subset of pandemic outbursts day or vice versa.

Therefore, the biclustering concept is used in this chapter as a pattern extraction technique for COVID-19 data that allows detecting groups of countries that cannot be found by a traditional clustering approach. Generally, the clustering approach always identifies a group of countries that exhibit the same epidemic trajectory pattern over the entire days of pandemic outbursts. Biclustering overcomes this limitation and groups the countries and the outburst days simultaneously into a bicluster. This improves the efficiency of the COVID-19 pattern analysis system.

In the literature, a different variation of biclustering algorithms has been proposed due to their ability to find collections of rows with a related behavior or pattern over a subgroup of columns [1,2]. In [3], Padilha V. A. *et al.* presented a comparative study involving 17 biclustering algorithms and concluded that the choice of the best biclustering algorithm purely depends on the types of patterns that are to be discovered and applied. In these review papers, "the authors described biclustering algorithms, their quality evaluation measures, and different types of validations used for the biclustering of gene expression data in detail" [4,5].

Let us consider the number of COVID-19 confirmed cases of seven countries for nine consecutive days (i.e., time points), which is given in Figure 1.1. A biclustering algorithm extracts the trajectory pattern by considering a subset of countries and days of the given two-dimensional COVID-19 data into a submatrix called bicluster. Here, identified bicluster indicates that countries 3–5 have the same COVID-19 epidemic trajectory pattern for the days D3–D6 which is highlighted in light gray color. However, a traditional clustering method would consider all the days (from D1 to D9) for grouping the similar countries. While countries 3–5 may not have similar

Countries	D1	D2	D3	D4	D5	D6	D7	D8	D9
1	V11	V12	V13	V14	V15	V16	V17	V18	V19
2	V21	V22	V23	V24	V25	V26	V27	V28	V29
3	V31	V32	V33	V34	V35	V36	V37	V38	V39
4	V41	V42	V43	V44	V45	V46	V47	V48	V49
5	V51	V52	V53	V54	V55	V56	V57	V58	V59
6	V61	V62	V63	V64	V65	V66	V67	V68	V69
7	V71	V72	V73	V74	V75	V76	V77	V78	V79

Figure 1.1 Sample dataset.

trajectory pattern for the entire period from D1 to D9, one can easily observe strong similarities for the period D3–D6. Therefore, countries 3–5 can be viewed as a bicluster for the days D3–D6 rather than as being clustered over the entire days with low similarity.

Harigan proposed the biclustering approach for the first time to explore voting data of US states [1]. Harigan used variance as a quality measure to identify the bicluster. Mean square residue (MSR) was introduced by Cheng and Church to assess the quality of the bicluster [4]. Teng L and Chan L. W [4] "defined a statistical score namely average correlation value (ACV) for evaluating a bicluster based on the weighted correlation coefficient." In [6], the authors reviewed the development of various biclustering algorithms in a comprehensive way.

1.2 Problem Description

In the literature, it has been found that many biclustering approaches or techniques have used some similarity measure for assessment of the bicluster quality as well as for extracting the bicluster [9]. Popular MSR measure is used very frequently in many biclustering algorithms. If elements in a bicluster or submatrix have less magnitude value, then the MSR is also low. In [5], "biclusters ought to contain a set of rows showing similar behavior, not similar values. A correlated subset of rows will demonstrate pure scaling or shifting patterns across a subset of columns. Such correlation is pattern-based, that is neither linear nor nonlinear."

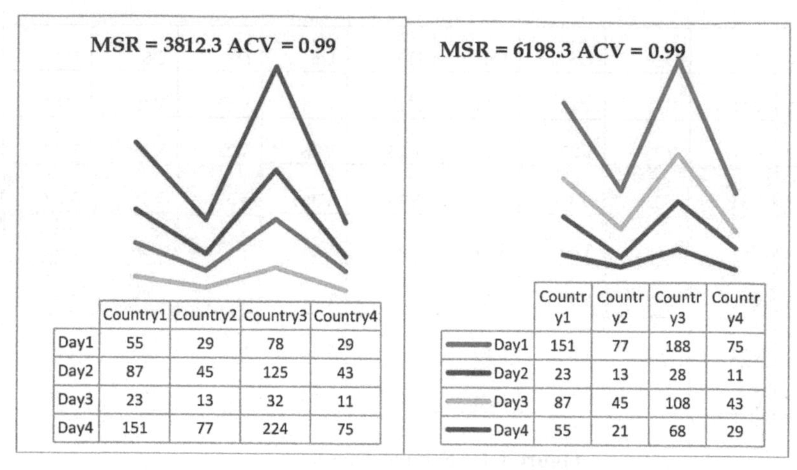

	Country1	Country2	Country3	Country4
Day1	55	29	78	29
Day2	87	45	125	43
Day3	23	13	32	11
Day4	151	77	224	75

	Country1	Country2	Country3	Country4
Day1	151	77	188	75
Day2	23	13	28	11
Day3	87	45	108	43
Day4	55	21	68	29

Figure 1.2 Similar bicluster pattern with different values.

Let us consider two biclusters in Figure 1.2 whose pandemic outburst days vary in unison under the countries contained in the bicluster. Despite the fact the countries exhibit the same epidemiological trajectory under the subset of outburst days, the MSR value for these two biclusters is different; there is a huge difference in their magnitude (value) which denotes they are not equally good biclusters. Comparing these two biclusters graphically, it cannot be concluded that the left-side bicluster is better than the right-side bicluster as it would be unfair to claim that only countries having less COVID-19 cases for certain days are preferable.

ACV, a bicluster quality measure, evaluates well all types of biclusters and also tolerates transformations like scaling and translation [7, 8, 9, 10]. The ACV of these two biclusters is 0.99. Therefore, MSR is not the best quality measure to discover bicluster when the variance of COVID-19 cases' values is high. Hence, in this research work, correlation-based merit function called ACV has been applied to extract the biclusters.

1.2.1 Greedy Approach: Bicluster Size Maximization Based Fitness Function

The greedy approach aims to find maximum size highly correlated biclusters having shifting and scaling patterns. "Hence therefore, a following correlation measure can be a good choice of fitness function to find larger size biclusters with these patterns" [7, 8, 9]. It can be represented as in Equation (1.1).

$$Dy = \alpha * Dx + \beta, \text{ where } \alpha, \beta \in R. \tag{1.1}$$

Let us consider B as bicluster composed by N days, $B = \{d1, ..., dN\}$, the correlation quality of B, $\rho(B)$, is defined in Equation (1.2) as follows:

$$\rho(B) = \frac{\sum_{i=1}^{N-1} \sum_{j=i+1}^{N} \rho(d_i, d_j)}{n^2 - n} \tag{1.2}$$

where $\rho(d_i, d_j)$ is the correlation coefficient between the day d_i and the user d_j which measures the linear grade dependency between X and Y variables. It is defined in Equation (1.3).

$$\rho(X, Y) = \frac{\sum_{i=1}^{n} (x_i - \mu x) - (y_i - \mu y)}{\sigma_X * \sigma_Y} \tag{1.3}$$

where μx and μy are the mean value of variables X and Y and σX and σY are the standard deviation values of X and Y, respectively. The fitness function defined in Equation (1.4) is used to assess the bicluster's quality subjected to given criteria.

$$\begin{aligned} \max \quad & f(D', C') = |D'| * |C'| \\ \text{subjected to} \quad & g(D', C') \end{aligned} \tag{1.4}$$

where B (D', C') is the bicluster, $g(D', C') = (1 - \rho(B))$, $|D'|$ and $|C'|$ are the number of pandemic outburst days and countries in the bicluster, respectively, and μ is the correlation threshold.

1.2.2 Data Description

COVID-19 data is used, which have downloaded from the Kaggle Website. Since January 22, 2020, the data is updated on a daily basis, of the increment of the number of infected people in COVID-19 infected countries across the world. It contains epidemiological data such as COVID-19 time series confirmed cases, COVID-19 time series recovered cases, and COVID-19 time series death cases, but COVID-19 time series new cases are derived from the above data. For this study, COVID-19 time series data for the above-mentioned cases have been taken for the period from March 11, 2020 to April 10, 2020. In this period, worldwide, millions of peoples are suffering from COVID-19 severely. Also, the epidemiological trajectory moved toward their new peak value in many countries.

The number of COVID-19 confirmed cases increases exponentially in most parts of the globe. Especially in the US, the pandemic is very severe and collapsing their healthcare

Pseudo code: Greedy Biclustering Algorithm

Initialize input data, *R* – set of rows in the data, *C* – set of columns in data
Step 1: **Data Preprocessing:** Convert COVID-19 epidemiological data into the data matrix for top 50 countries.
Step 2: **Pattern Identification:**
Rowcluster(Data,rK)//Apply KMeans clustering algorithms on rows of the "Data"
Columncluster(Data,cK)//Apply KMeans clustering algorithms on columns of the "Data"
Combine(Rowcluster, Columncluster)// to form bicluster
Step 3: **Optimize the Biclusters**
For each "*B*"
Enlarge (*B*)
Refine(*B*)
For end
Step 3.1: sub **Enlarge()**
br = set of rows in *B*
br = set of columns in *B*
r = setdifference(*R*,*br*)
c = setdifference(*C*,*bc*)
For each node *(row/column)*
If ACV(union(*Seed, r/c*)) > ACV(Seed(*R, C*)) then
Add *r/c* to Seed(*R*,*C*)
End(if)
End(for)
End sub
Step 3.2: sub **Refine()**
For each node *r/c* in Enlarged Seed
Remove node *r/c* from Enlarged Seed
R″/C″ be set of rows/columns in R′/*C′* but not contained *r/c*
If ACV (Enlarged Seed(*R″*,*C″*)) >
ACV(Enlarged Seed(R′, *C′*)
Update R′/*C′*
End(if)
End(for)
End sub
Step 4: **Pattern analysis:** Post process the result

system [11, 12]. If these pandemic trends continue, it will affect other countries in a short period of time badly. Therefore, a reliable and accurate analysis and forecasting system for COVID-19 spread are required to contain the novel coronavirus spread [13-23].

In [17], the authors presented a simple prediction method to forecast the number of COVID-19 cases and the insights and strategies on basic principles for COVID-19 are mentioned. According to our knowledge, there is no work yet for pattern analysis of COVID-19 epidemiological data. Therefore, local patterns (i.e., biclusters) are extracted for top 50 COVID-19 infected countries using Greedy Biclustering Algorithms in this work.

1.3 Proposed Work: COVID 19 Pattern Identification Using Greedy Biclustering

Given a COVID-19 data matrix D, let n_r be the number of row (day) clusters and n_c be the number of column (country) clusters. After K-Means clustering [24, 25] is applied on both row and column dimensions, CC^r is the group of row (day) clusters and CC^c is the group of column (country) clusters. Let ${cc_i}^r$ be a subgroup of outburst days and ${cc_i}^r \in C^r$ $(1 \leq i \leq k_r)$. Let ${cc_j}^c$ be a subgroup of countries and ${cc_j}^c \in C^c$ $(1 \leq j \leq k_c)$. The pair $({cc_i}^r, {cc_j}^c)$ represents a bicluster of D. The results of row dimensional clustering and column dimensional clustering are merged to obtain $n_r \times n_c$ biclusters. These biclusters are called seeds [8, 26-30].

Optimize Biclusters Using Greedy Approach:

In this step, the seeds are extended and refined using the Greedy approach. This process adds/removes the rows (outburst days) and columns (countries) into/from the existing bicluster for increasing their volume/improving their ACV, respectively [7, 8, 31]. The purpose of using Greedy heuristic is to increase the volume of the bicluster (seed) without degrading its quality. The volume of bicluster is defined as the multiplication of number of countries and days in that bicluster [8]. Here, ACV is used as quality criteria to grow the seeds. In this Greedy Biclustering approach, one can increase the ACV of the bicluster by inserting/deleting the rows (days)/columns (countries) to/from the given initial bicluster. This process is also known as seed enlargement and refinement.

Table 1.1 Characteristics of biclusters for COVID-19 death cases.

Bicluster index	No. country	No. of days	ACV	Volume
1	37	16	0.99	592
2	37	9	0.99	333
3	37	6	0.99	222
4	8	16	0.99	128
5	13	9	0.99	117
6	15	6	0.99	90
7	4	16	0.99	64
8	4	31	1	124

Table 1.2 Characteristics of biclusters for COVID-19 recovered cases.

Bicluster index	No. country	No. of days	ACV	Volume
1	42	19	0.99	798
2	41	8	0.99	328
3	42	4	0.99	168
4	7	19	0.99	133
5	5	19	0.99	95
6	18	4	0.99	72
7	5	31	1	155

1.4 Results and Discussions

In this section, the results of the most representative bicluster obtained on the COVID-19 data are presented and summarized. Tables 1.1 and 1.2 show the characteristics of extracted biclusters for top 50 COVID-19 infected countries' death cases and recovered cases data, respectively. These tables contain the following:

- the number of countries in the biclusters;
- the number of the time period (days) in the biclusters;
- quality and quantitative measure: ACV of the biclusters and the volume of the biclusters.

These characteristic information of extracted biclusters provide us an idea of how epidemiological trajectory behaves similarly among the subsets of countries over the subset of days taken for the study. From the presented tables, it can be easily noted that all identified biclusters are highly correlated since their ACV is almost equal to value 1. It indicates that the Greedy Biclustering approach captured the highly correlated bicluster very well (i.e., biclusters with scaling and shifting pattern).

It has been observed from Figures 1.3(a–g) that:

- As of April 10, 2020, the United States topped the COVID-19 mortality rate and Italy was second. The US, Italy, Spain, and Iran have higher mortality rates compared to other countries.

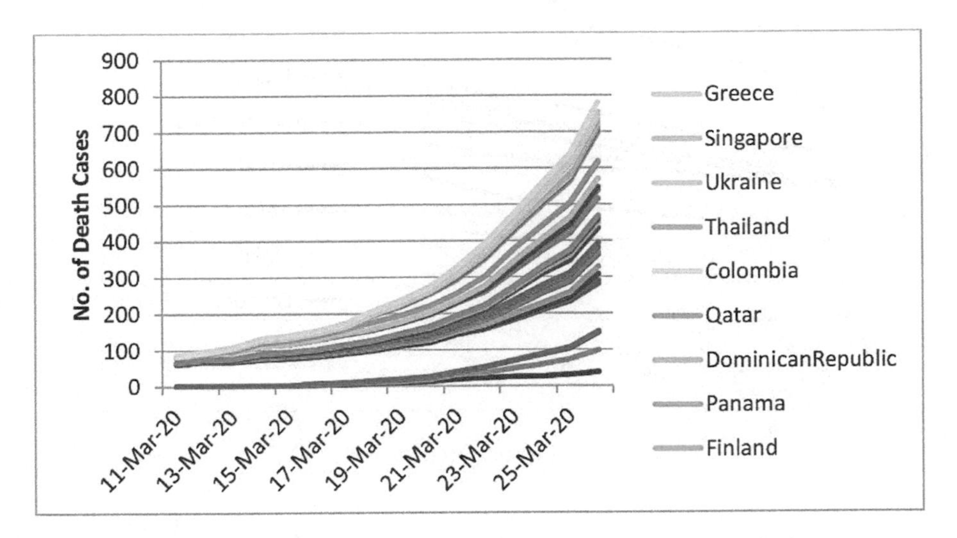

Figure 1.3(a).

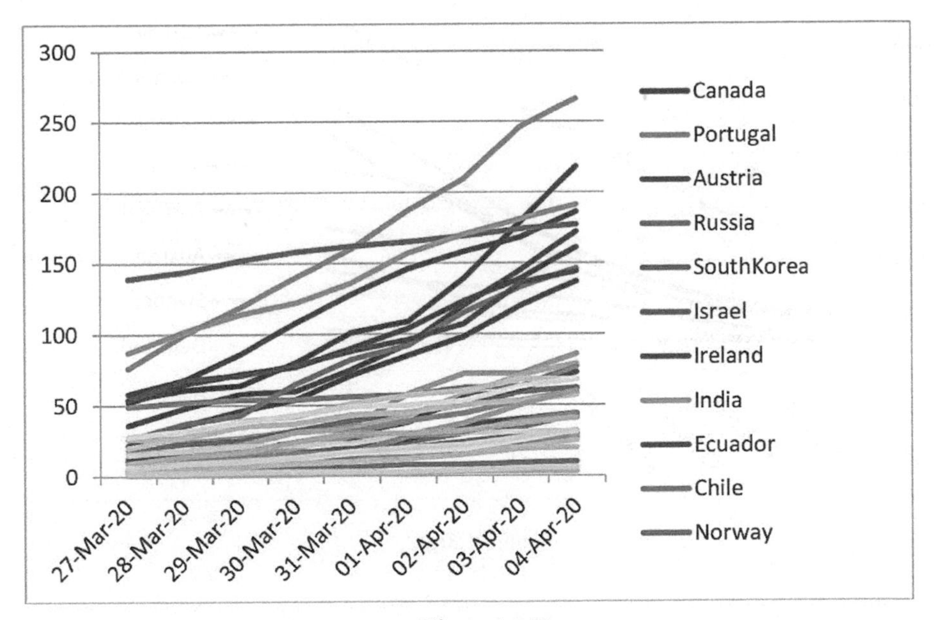

Figure 1.3(b).

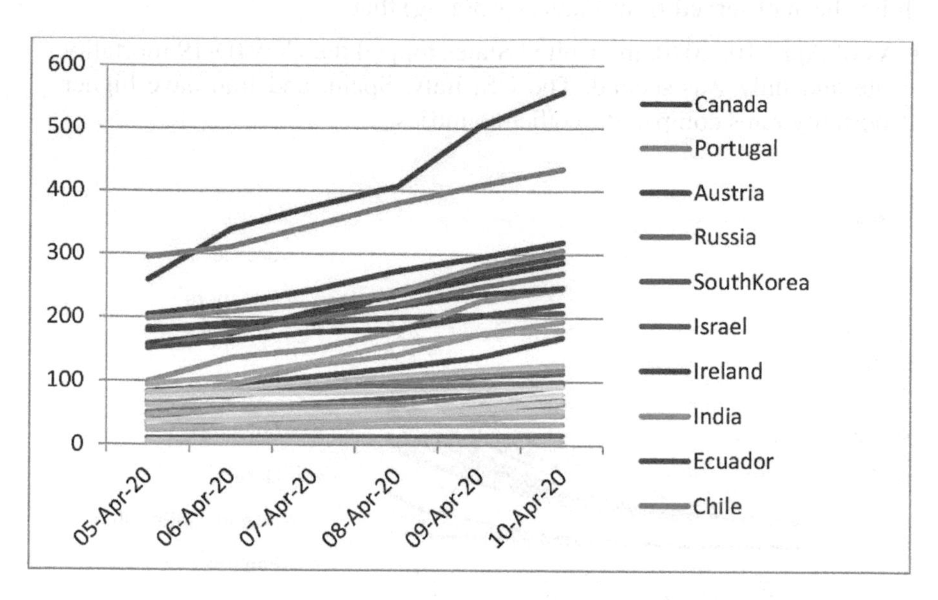

Figure 1.3(c).

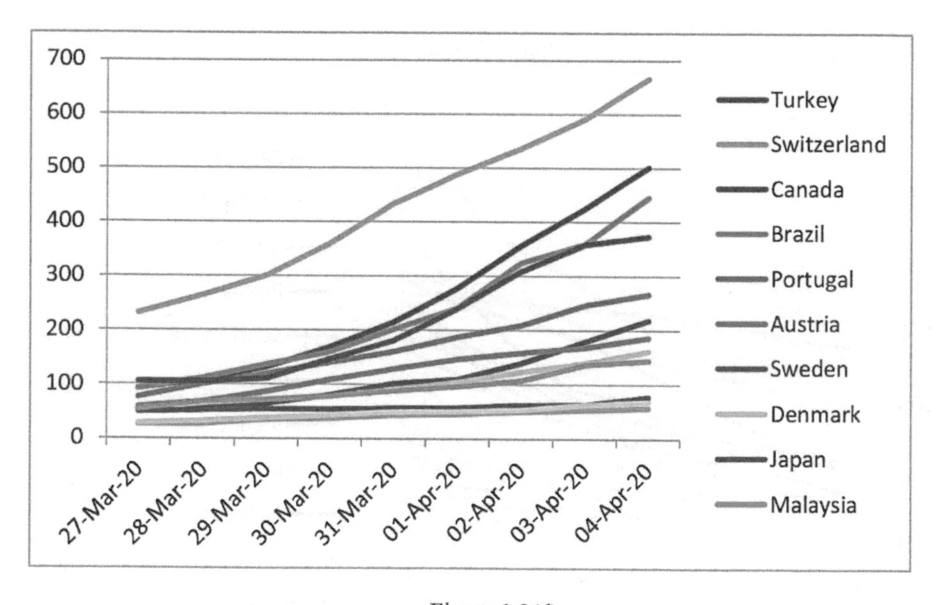

Figure 1.3(d).

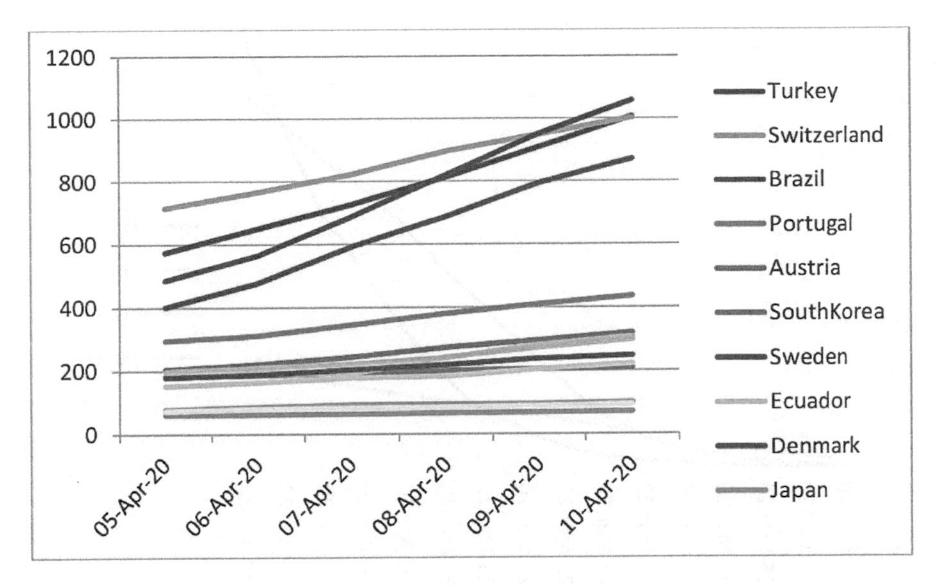

Figure 1.3(e).

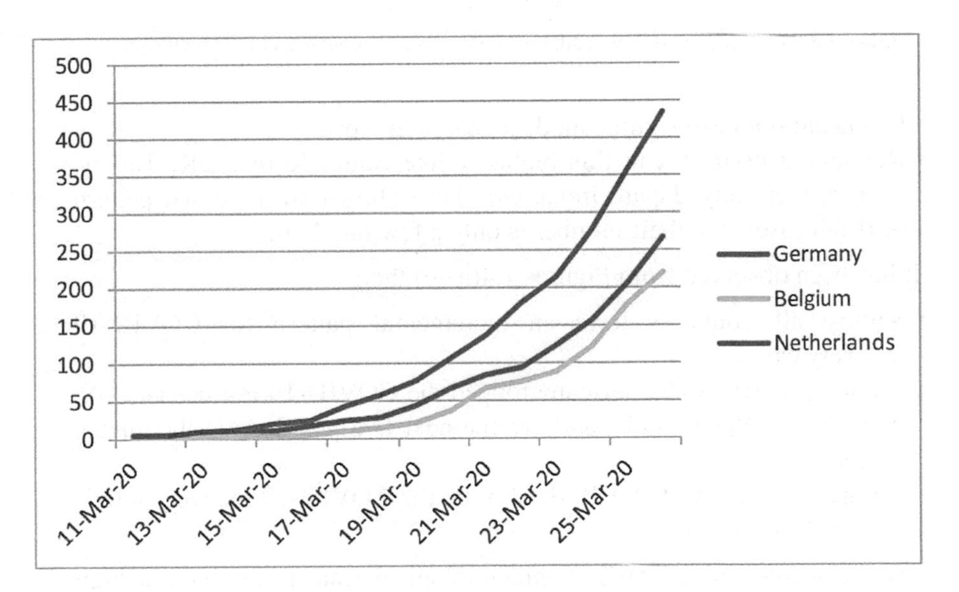

Figure 1.3(f).

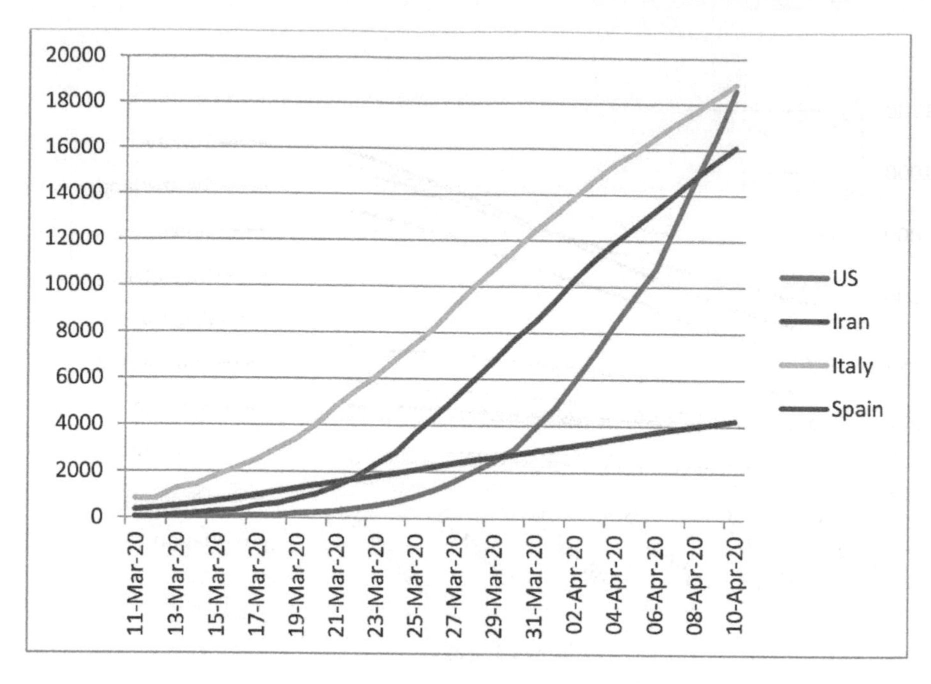

Figure 1.3(g).

Figure 1.3 (a–g) COVID-19 death cases patterns over countries in the bicluster.

- China did not have significant death cases in April.
- Remaining countries in this bicluster like South Korea, UK, Turkey, France, Germany, Japan, India, etc., have shown an increased pattern of death cases, but their number is only a few hundred.

It has been observed from Figures 1.4(a–e) that:

- Almost all countries have an incremental pattern for COVID-19 recovery cases.
- As of April 10, 2020, Germany topped the COVID-19 rescue. The US, Italy, China, Spain, and Iran have the next five positions in the queue, respectively.
- Compared to the COVID-19 death rate, the COVID-19 recovery rate is low in all countries.

On the whole, all countries' epidemiological trajectories have a high correlation with the US. Therefore, rapid and reliable measures, such as social

Table 1.3 Description of Figure 1.3(a–g).

Figure	Countries in the Bicluster	Remarks
1.3(a)	Canada, Portugal, Austria, Russia, South Korea, Israel, Ireland, India, Ecuador, Chile, Norway, Australia, Denmark, Poland, Peru, Czech Republic, Japan, Romania, Pakistan, Malaysia, Philippines, Saudi, Indonesia, Mexico, UAE, Luxembourg, Serbia, Finland, Panama, Dominican Republic, Qatar, Colombia, Thailand, Ukraine, Singapore, and Greece	COVID-19 death cases for these countries have shown increased patterns from March 11 to March 20. In particular, there are 60 plus daily COVID-19 death cases in Canada during this period. The maximum value for other countries is below 60.
1.3(b)	US, Canada, Portugal, Austria, Russia, South Korea, Israel, Ireland, India, Ecuador, Chile, Norway, Australia, Denmark, Poland, Peru, Czech Republic, Japan, Romania, Pakistan, Malaysia, Philippines, Saudi, Indonesia, Mexico, UAE, Luxembourg, Serbia, Finland, Panama, Dominican Republic, Qatar, Colombia, Thailand, Ukraine, Singapore, and Greece	Time period: March 7 to April 4, 2020. The number of death cases for the US has increased by 400 plus daily, with a value of approximately 1500–8000; countries like Canada, Portugal, Austria, and South Korea have a daily increase of death cases by 20 plus cases; and remaining countries have a daily increase of 10 cases.
1.3(c)	US, Canada, Portugal, Austria, Russia, South Korea, Israel, Ireland, India, Ecuador, Chile, Norway, Australia, Denmark, Poland, Peru, Czech Republic, Japan, Romania, Pakistan, Malaysia, Philippines, Saudi, Indonesia, Mexico, UAE, Luxembourg, Serbia, Finland, Panama, Dominican Republic, Qatar, Colombia, Thailand, Ukraine, Singapore, and Greece	Time period: April 5 to April 10, 2020. In the US, there were about 2000 cases of daily deaths. In the rest of the country, there were about 100 cases of daily deaths.
1.3(d)	Turkey, Switzerland, Canada, Brazil, Portugal, Austria, Sweden, Denmark, Japan, Malaysia, Philippines, and Greece	Time period: March 27 to April 4, 2020. Daily death cases for these countries have increased by approximately 40–50 cases
1.3(e)	Turkey, Switzerland, Brazil, Portugal, Austria, South Korea, Sweden, Ecuador, Denmark, Japan, Malaysia, Philippines, Indonesia, and Greece	Time period: April 5 to April 10, 2020. Daily death cases for countries Turkey, Belgium, Switzerland, Netherlands, Canada, and Portugal have increased by 30–50 cases approximately, whereas that for the remaining countries have increased by 10–20 cases daily.

Table 1.3 *Continued*

1.3(f)	Germany, Belgium, and Netherland	Time period: March 11 to March 26, 2020. In the early days, the number of cases of daily death was very low, but it has gradually increased by 30–50 daily cases.
1.3(g)	US, Iran, Italy, and Spain	Time period: March 11 to April 10, 2020, In the early days, the number of cases of daily death was very low, but it has gradually increased by a few hundred to a few thousand daily. As of April 10, 2020, the US topped the COVID-19 mortality rate which is followed by Italy and Spain.

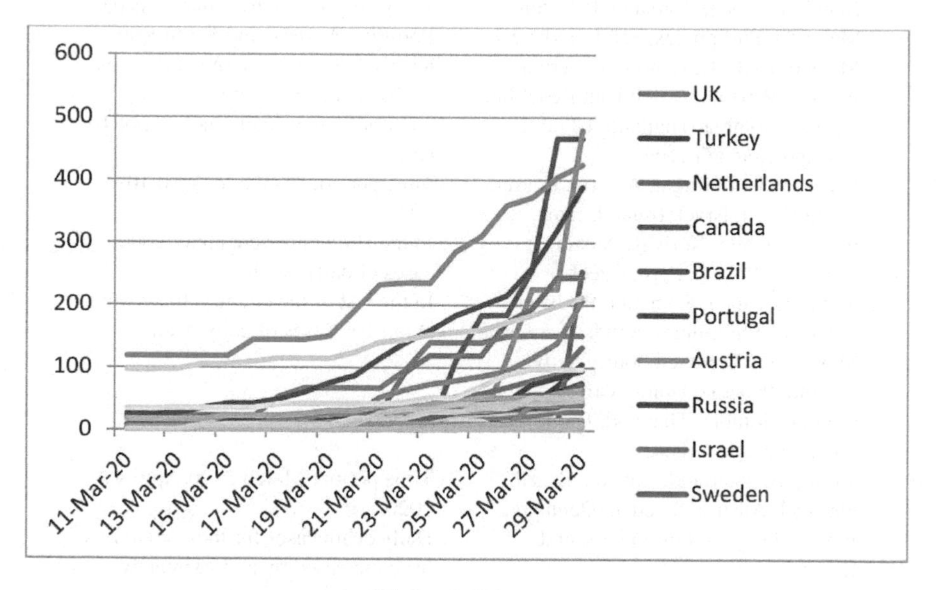

Figure 1.4(a).

distance, health practices, and national lockdown, are needed to control the spread of the coronavirus. Otherwise, this virus is likely to spread so badly in all countries, like the United States.

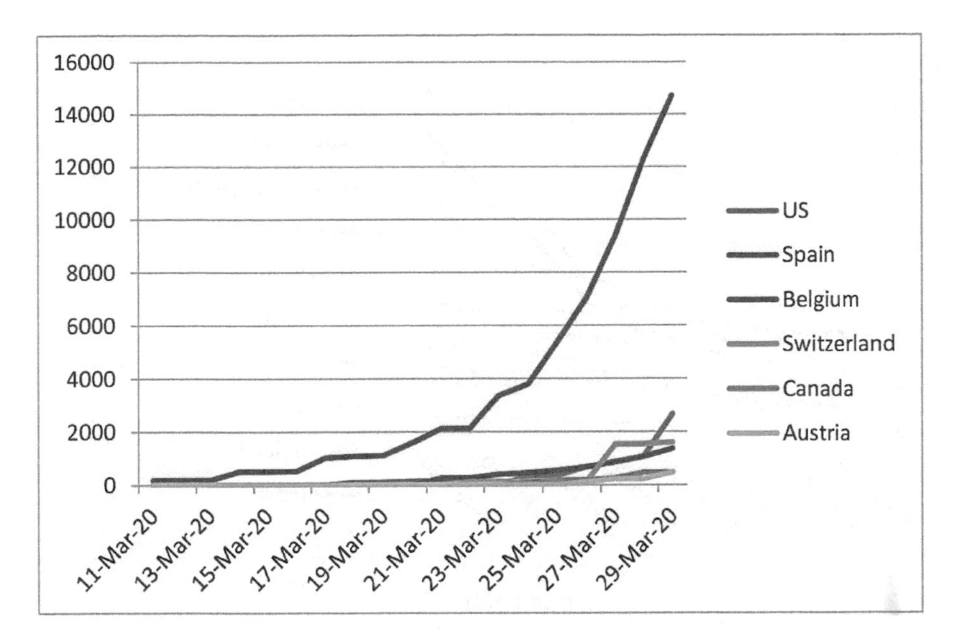

Figure 1.4(b).

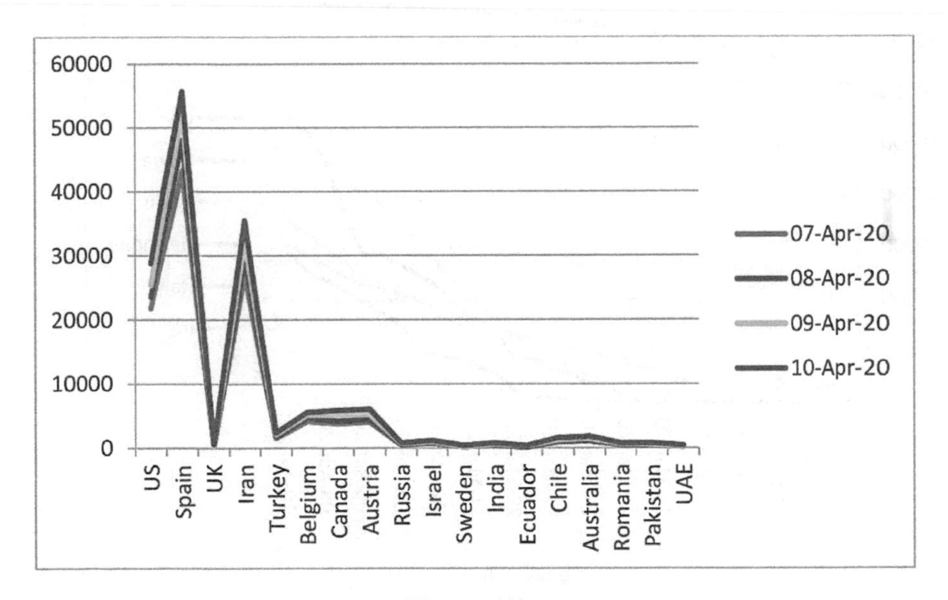

Figure 1.4(c).

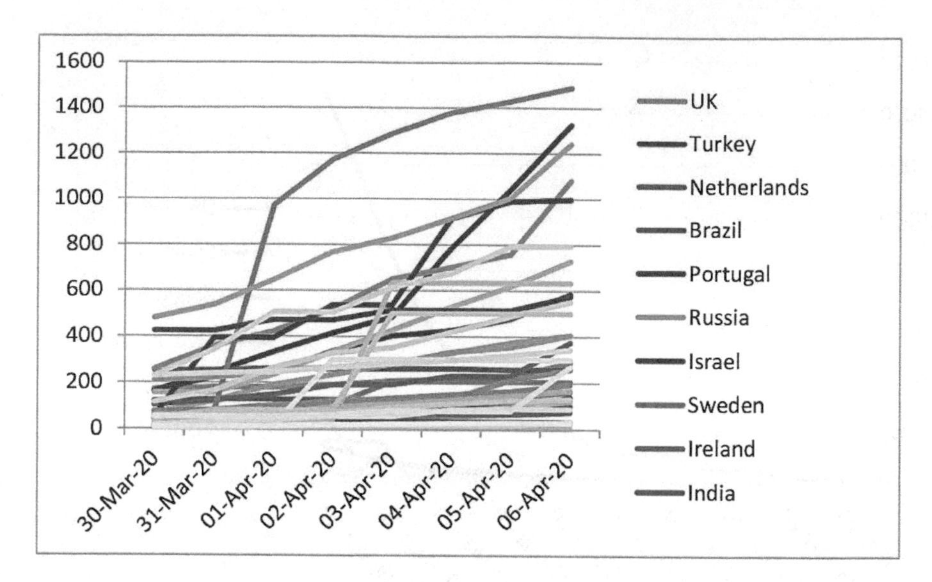

Figure 1.4(d).

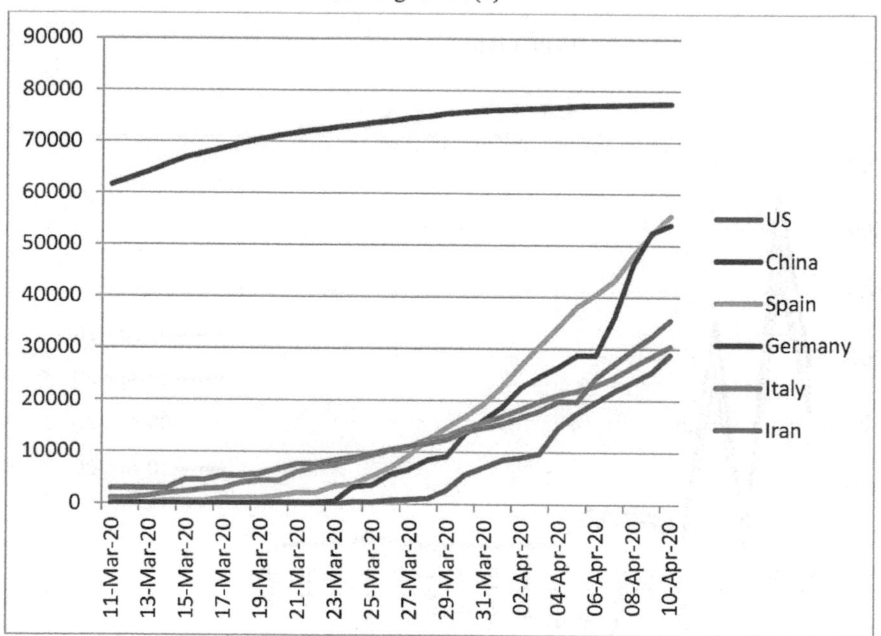

Figure 1.4(e).

Figure 1.4 (a–e): COVID-19 recovered cases pattern for countries in the bicluster.

Table 1.4 Description of Figure 1.4(a–e).

Figure	Countries in the bicluster	Remarks
1.4(a)	UK, Turkey, Netherlands, Canada, Brazil, Portugal, Austria, Russia, Israel, Sweden, Ireland, India, Ecuador, Chile, Norway, Australia, Denmark, Poland, Peru, Czech Republic, Japan, Romania, Pakistan, Malaysia, Philippines, Saudi, Indonesia, Mexico, UAE, Luxembourg, Serbia, Finland, Panama, Dominican Republic, Qatar, Colombia, Thailand, Ukraine, Singapore, and Greece	COVID-19 recovered cases for these countries have shown increased pattern from March 11 to March 29, 2020. In particular, countries like Turkey, Brazil, Netherland, Ireland, Chile, Peru, etc., have no COVID-19 recovery cases in the second week of March 2020. But gradually, it has risen to 100 in all countries.
1.4(b)	US, Spain, China, Belgium, Switzerland, Canada, and Austria	COVID-19 recovered cases for these countries have shown increased pattern from March 11 to March 29, 2020. In the early days, one or two cases have recovered from COVID-19, but by the end of March 2020, it has risen to 1000 plus in these countries.
1.4(c)	US, Spain, UK, Iran, Turkey, Belgium, Canada, Austria, Russia, Israel, Sweden, India, Ecuador, Chile, Australia, Romania, Pakistan, and UAE	Time period: April 7 to April 10, 2020. For the US, Spain, and Iran, daily recovered cases have increased to 2000 plus approximately, whereas in the remaining countries, daily recovered cases have increased to 100 plus approximately.
1.4(d)	UK, Turkey, Belgium, Netherlands, Brazil, Portugal, Russia, Israel, Sweden, Ireland, India, Ecuador, Chile, Norway, Australia, Denmark, Poland, Peru, Czech Republic, Japan, Romania, Pakistan, Malaysia, Philippines, Saudi, Indonesia, Mexico, UAE, Luxembourg, Serbia, Finland, Panama, Dominican Republic, Qatar, Colombia, Thailand, Ukraine, Singapore, and Greece	Time period: March 30 to April 6, 2020. Daily recovered cases for these countries have increased from 30 to 100 cases approximately
1.4(e)	US, China, Spain, Germany, Italy, and Iran	Time period: March 11 to April 10, 2020 In the early days, the number of daily recovered cases was very low, but it has gradually increased from a few hundred to a few thousand daily.

1.5 Conclusion

In this work, an analysis based on the biclustering of COVID-19 epidemiological data has been presented for the top 50 COVID-19 infected countries. Apparently, the Greedy Biclustering approach to COVID-19 data analysis has not been used until now. The main objective of this study was to detect meaningful and reliable patterns in a more optimized way. Instead of clustering, biclustering techniques allow discovering a subset of countries showing similar epidemic trajectory over a specific period (i.e., pandemic outburst days). This work has shown how biclustering methods could fit to study COVID-19 epidemiological time series data. The results obtained by the proposed approach will help public health officials to understand country-wise how COVID-19 is transmitted in humans, recovery rates, and death rates. In future, optimized biclustering approaches like evolutionary algorithms and swarm intelligence based biclustering approached will be implemented to analyze COVID-19 data.

1.6 Acknowledgements

The first author acknowledges the UGC-Special Assistance Programme (SAP) for the financial support to her research under the UGC-SAP at the level of DRS-II (Ref. No. F.5-6/2018/DRS-II (SAP-II)), July 26, 2018 in the Department of Computer Science.

References

[1] International Hartigan, J. A. (1972). Direct Clustering of a Data Matrix. Journal of the American Statistical Association, 67(337), 123–129. doi:10.1080/01621459.1972.10481214

[2] Cheng Y, Church GM. Biclustering of expression data. Proc Int Conf Intell Syst Mol Biol. 2000;8:93-103.

[3] Padilha, V. A., & Campello, R. J. G. B. (2017). A systematic comparative evaluation of biclustering techniques. BMC Bioinformatics, 18(1). doi:10.1186/s12859-017-1487-1

[4] Teng, L., & Chan, L. (2007). Discovering Biclusters by Iteratively Sorting with Weighted Correlation Coefficient in Gene Expression Data. Journal of Signal Processing Systems, 50(3), 267–280. doi:10.1007/s11265-007-0121-2

[5] Aguilar-Ruiz, J. S. (2005). Shifting and scaling patterns from gene expression data. Bioinformatics, 21(20), 3840–3845. doi:10.1093/bioinformatics/bti641

[6] Pontes B, Giráldez R, Aguilar-Ruiz JS. Biclustering on expression data: A review. J Biomed Inform. 2015;57:163-180. doi:10.1016/j.jbi.2015.06.028

[7] Dharan, S., & Nair, A. S. (2009). Biclustering of gene expression data using reactive greedy randomized adaptive search procedure. BMC Bioinformatics, 10(S1). doi:10.1186/1471-2105-10-s1-s27

[8] Rathipriya, R., Thangavel, K., & Bagyamani, J. (2011). Binary Particle Swarm Optimization based Biclustering of Web Usage Data. International Journal of Computer Applications, 25(2), 43–49. doi:10.5120/3001-4036

[9] Anitha, S., & Chandran, D. (2016). Review on Analysis of Gene Expression Data Using Biclustering Approaches. Bonfring International Journal of Data Mining, 6(2), 16–23. doi:10.9756/bijdm.8135

[10] Biswal, B. S., Mohapatra, A., & Vipsita, S. (2018). A review on biclustering of gene expression microarray data: algorithms, effective measures and validations. International Journal of Data Mining and Bioinformatics, 21(3), 230. doi:10.1504/ijdmb.2018.097683

[11] COVID-19 and Italy: what next? - The Lancet. https://www.thelancet.com/journals/lancet/article/PIIS0140-6736(20)30627-9/fulltext. Accessed 5 July 2020

[12] Li, Q., Guan, X., Wu, P., Wang, X., Zhou, L., Tong, Y., et al. (2020). Early Transmission Dynamics in Wuhan, China, of Novel Coronavirus–Infected Pneumonia. New England Journal of Medicine, 382(13), 1199–1207. doi:10.1056/nejmoa2001316

[13] Zhou, T., Liu, Q., Yang, Z., Liao, J., Yang, K., Bai, W., et al. (2020). Preliminary prediction of the basic reproduction number of the Wuhan novel coronavirus 2019-nCoV. Journal of Evidence-Based Medicine, 13(1), 3–7. doi:10.1111/jebm.12376

[14] Zhao, S., Lin, Q., Ran, J., Musa, S. S., Yang, G., Wang, W., et al. (2020). Preliminary estimation of the basic reproduction number of novel coronavirus (2019-nCoV) in China, from 2019 to 2020: A data-driven analysis in the early phase of the outbreak. International Journal of Infectious Diseases, 92, 214–217. doi:10.1016/j.ijid.2020.01.050

[15] Liu, Y., Gayle, A. A., Wilder-Smith, A., & Rocklöv, J. (2020). The reproductive number of COVID-19 is higher compared to SARS coronavirus. Journal of Travel Medicine, 27(2). doi:10.1093/jtm/taaa021

[16] Lai, A., Bergna, A., Acciarri, C., Galli, M., & Zehender, G. (2020). Early phylogenetic estimate of the effective reproduction number of SARS-CoV-2. Journal of Medical Virology, 92(6), 675–679. doi:10.1002/jmv.25723

[17] Perc, M., Miksić, N. G., Slavinec, M., & Stožer, A. (2020). Forecasting COVID-19. Frontiers in Physics, 8. doi:10.3389/fphy.2020.00127

[18] Dhamodharavadhani, S., & Rathipriya, R. (2021). COVID-19 mortality rate prediction for India using statistical neural networks and gaussian process regression model, African Health Sciences 21 (1), 194-206.

[19] Namasudra, S., Dhamodharavadhani, S., & Rathipriya, R. (2021). Nonlinear Neural Network Based Forecasting Model for Predicting COVID-19 Cases. Neural Process Lett. https://doi.org/10.1007/s11063-021-10495-w

[20] Dhamodharavadhani, S., Rathipriya, R., & Chatterjee, J. M. (2020). COVID-19 Mortality Rate Prediction for India Using Statistical Neural Network Models. Front Public Health. 8:441. Published 2020 Aug 28. doi:10.3389/fpubh.2020.00441

[21] Devipriya, R., Dhamodharavadhani, S., & Selvi, S. (2021). SEIR model FOR COVID-19 Epidemic using DELAY differential equation. Journal of Physics: Conference Series, vol. 1767, no. 1, p. 012005.

[22] Dhamodharavadhani, S., & Rathipriya, R. (2021). Novel COVID-19 Mortality Rate Prediction (MRP) Model for India Using Regression Model With Optimized Hyperparameter. Journal of Cases on Information Technology, vol. 23, no. 4, pp. 1-12.

[23] Dhamodharavadhani, S., & Rathipriya, R. (2018). Region-Wise Rainfall Prediction Using MapReduce-Based Exponential Smoothing Techniques. Advances in Intelligent Systems and Computing, pp. 229-239.

[24] Dhamodharavadhani, S., & Rathipriya, R. (2020). Enhanced Logistic Regression (ELR) Model for Big Data. Handbook of Research on Big Data Clustering and Machine Learning Advances in Data Mining and Database Management, 152–176. doi:10.4018/978-1-7998-0106-1.ch008

[25] Dhamodharavadhani, S., & Rathipriya, R. (2020). Variable Selection Method for Regression Models Using Computational Intelligence Techniques. Handbook of Research on Machine and Deep Learning Applications for Cyber Security Advances in Information Security, Privacy, and Ethics, 416–436. doi:10.4018/978-1-5225-9611-0.ch019

[26] Dhamodharavadhani, S., & Rathipriya, R. (2020). Forecasting Dengue Incidence Rate in Tamil Nadu Using ARIMA Time Series Model. Machine Learning for Healthcare, pp. 187-202.

[27] Sivabalan, S., Dhamodharavadhani, S., & Rathipriya, R. (2020). Arbitrary walk with minimum length based route identification scheme in graph structure for opportunistic wireless sensor network. Swarm Intelligence for Resource Management in Internet of Things, pp. 47-63.

[28] Sivabalan, S., Dhamodharavadhani, S., & Rathipriya, R. (2019). pportunistic Forward Routing Using Bee Colony Optimization. International Journal of Computer Sciences and Engineering, vol. 7, no. 5, pp. 1820-1827.

[29] Kaleeswaran, V., Dhamodharavadhani, S., & Rathipriya, R. (2020). A Comparative Study of Activation Functions and Training Algorithm of NAR Neural Network for Crop Prediction. 2020 4th International Conference on Electronics, Communication and Aerospace Technology (ICECA).

[30] Kaleeswaran, V., Dhamodharavadhani, S., & Rathipriya, R. (2021). Multi-crop Selection Model Using Binary Particle Swarm Optimization. Innovative Data Communication Technologies and Application, pp. 57-68.

[31] Das, S., & Idicula, S. M. (2010). Greedy Search-Binary PSO Hybrid for Biclustering Gene Expression Data. International Journal of Computer Applications, 2(3), 1–5. doi:10.5120/651-908

2

Artificial Fish Swarm Optimization Algorithm with Hill Climbing Based Clustering Technique for Throughput Maximization in Wireless Multimedia Sensor Network

N. Krishnaraj[1,*], T. Jayasankar[2], N. V. Kousik[3] and A. Daniel[4]

[1]Associate Professor, School of Computing, SRM Institute of Science and Technology, Kattankulathur 603203, Tamilnadu, India
[2]Assistant Professor, Department of ECE, University College of Engineering, BIT Campus, Anna University, Tiruchirappalli-620024, Tamilnadu, India.
[3]Associate Professor, School of Computing Science and Engineering, Galgotias University, Greater Noida, Uttar Pradesh, India.
[4]Assistant Professor, School of Computing Science and Engineering, Galgotias University, Greater Noida, Uttar Pradesh, India.
E-mail: [1]drnkrishanraj@gmail.com [2]jayasankar27681@gmail.com, [3]nvkousik@gmail.com, [4]danielarockiam@gmail.com
*Corresponding Author: N. Krishnaraj, drnkrishnaraj@gmail.com

Abstract

Wireless multimedia sensor networks (WMSNs) offer useful information in the tracking and surveillance applications, which processes scalar data, images, audios, and videos. But the multimedia streaming is highly demanding for the networks, as energy constraint sensor nodes restrict the possible bandwidth for data transmission and leads to reduced throughput. Energy efficiency and throughput maximization are the two key design issues of WMSN, which can be attained by the clustering process. The use of clustering technique helps to group the sensor nodes into clusters, and a cluster head (CH) will be elected among every cluster. In this view, this paper presents a new artificial fish swarm optimization algorithm (AFSA) with hill climbing (HC) based clustering technique for throughput maximization in WMSN, called AFSA-HC. The proposed AFSA-HC algorithm involves four main processes to maximize throughput in the network, namely node initialization, AFSA-HC based node clustering, deflate algorithm based data aggregation, and hybrid data transmission. A comprehensive experimental analysis is carried out to verify the effective performance of the presented AFSA-HC technique. The simulation outcome inferred that the AFSA-HC technique has attained maximum results in terms of different measures, namely energy consumption, throughput, network lifetime, network stability, and packet loss.

Keywords: Clustering, data aggregation, energy efficiency, throughput, wireless multimedia sensor network (WMSN).

2.1 Introduction

Currently, wireless multimedia sensor network (WMSN) used *t* for observing and managing a huge set of applications in different circumstances like visual surveillance, healthcare assistance, traffic operations, public safety, industrial automation, etc. Because of reduction in cost of wireless and sensor devices, WMSN is architected to work in diverse network backbone, merging Internet services by application-oriented networks [1]. Internet of Things (IoT) raises the domain of concurrent communications in which the conventional Internet principles might be integrated into the wireless sensor networks scenario.

WMSNs have possibilities of framing countless source nodes which streams data continuously or particular time period. The basic architecture of WMSN is shown in Figure 2.1. In this case, there might be many data that would stream over the network in the direction of sink. For *ad hoc* network consisting of wireless personal area mesh topologies [2], huge quantity of packets might not be sent because of given less bandwidth. Likewise, numerous packets can be delivered via *ad hoc* paths at the same time; it consumes energy in the nodes. In case of large-scale WMSN with more number of lively multimedia sensor nodes, medium access control (MAC) protocols for maximum bandwidth needs to be monitored, where IEEE 802.11 network becomes low-cost and appropriate situation. When

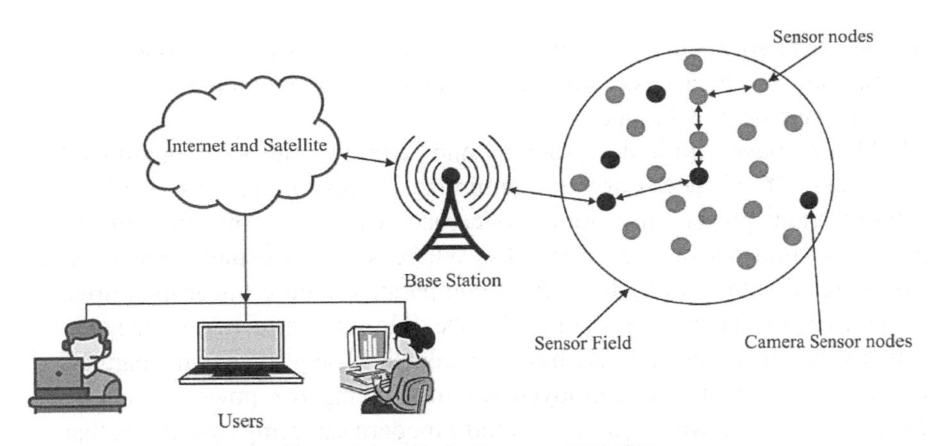

Figure 2.1 Block diagram of WMSN.

Wi-Fi access points (APs) would perform the sink responsibility, then higher bandwidth is necessary for WMSN.

It is essential to underline that the wireless communications raise quite a few effects comprising that of energy usage, which is related to Wi-Fi and WMSN context equally. Nowadays, there are lots of laborious works that directly aim to fix issues by launching wireless communication system with less power consumption, and, globally, the standard usage of energy network apparatus has been calculated to be 25 GW in day-to-day activities, and a single community wireless local area network (WLAN) might feature higher than 5000 Wi-Fi APS. However, as exposed in [3], these ranges have arisen during the past years. At the same time, regarding the current network designing technique, it can be helpful to notice that there is an absence of many essential components. Primarily, the traffic demands and user density are ignored. The actual coverage-based optimization technique might look inadequate for network where compactness of the consumer and traffic load is heavy.

WLAN and WMSN background would have high user attention very soon, with increase in application usage data value, but energy capacity would be still a major design concern. In fact, multimedia can possess unique transmission prototypes which alter overtime period. Camera-enabled sensing gadgets are supposed to send images or video stream with various broadband needs. Scalar sensing gadgets might be staked to recover the limited volume of data. Further, the source node might transfer data packets at the instance, demanding minimum latency and jitter [4]. For the purpose of supplying an energy-efficient solution among several sensor nodes associated with one-hop Wi-Fi access ends, this research work recommends a new energy management technique.

WMSNs have limited resource and exhibit elevated broadband requirement. It is necessary to minimize computational as well as communication power utilization concerned with data transmission to enhance the lifespan of such networks. Whereas, CS is usually employed in enabling all signals to be fixed from comparatively little linear measures. It is familiar for catching and portraying flexible signs at a value under the Nyquist value. It senses and may have the value at base intricacy in a parallel way. It reports that CS is employed for minimizing the power saving in WMSN. Candes and Wakin [5] discovered a modern sampling hypothesis that merges sampling and compression processes at the time of data collection. The usage of CS hypothesis could retrieve less sign and images from remote lesser samples or measures compared to the conventional techniques in the WMSN currently [6]. The sparse vector could be used for roughly calculating numerical data and useless measures by CS. Moreover, it could decrease the power utilization of the image communication task.

Multimedia packets could be sent via diverse wireless methodology, but quality of service (QoS) limitations such as delay and jitter must be reviewed all the time when it is mandatory by applications. Generally, a number of tasks have suggested a development tried at elevated achievement. In wireless sensing meshes, streaming multimedia inflict several difficulties which are known to be discussed in various tasks. The efforts in [7] offered the reliable synchronous transport protocol (RSTP) for synchronization of image movement from many source of sensing devices. Whereas, progressive coding is utilized for image processing [8]. However, images that were collected with bad standards could be posted completely when streamed along low-bandwidth nodes and erroneous medium.

In [9], journalists planned a computerized result to reduce blocks by lowering the data transmitting rate using the discrete wavelet transform (DWT) coding approach; the quality of images is not highly decreased. In various types, the effort in [10] defines a cross-layer optimization of video transmitting in WMSN. Researches employed multipath routing, where paths with less latency must be exploited frequently for transmitting data packets including encoded video and power consumption and may be possible to extend the capacity of the network by liberating real-time multimedia passage.

In [11], a cross-layer technique for streaming videos in WMSN was planned, and the idea of compressive sensor was employed, targeting the alteration for rates of data transmission. For all, when several network sources stream premium video, the broadband needs might be spectacular for sensing nodes. For helping good multimedia streaming in wireless sensing nodes, IEEE 802.11 MAC protocols are developed. A few investigations have discussed the issues of Wi-Fi based WMSN. In [12], journalists developed a difference in handling of multimedia data packets in MAC layer based on the importance of the encoded data for MPEG-4 video streaming. It exploits IEEE 802.11s extension [13], which supplies QoS-based transmission classes. In another way, in [14], the IEEE 802.11e extension is addressed for wireless multimedia sensing networks and journalists survived several MAC access control machinery of this protocol [15]. However, usually power consumption is unconcerned. For heavy wireless multimedia, sensor networks and power preservation must be examined attentively, promoting creative research in this domain.

This paper introduces a new artificial fish swarm optimization algorithm (AFSA) with hill climbing (HC) based clustering technique for throughput maximization in WMSN, called AFSA-HC. Since the traditional AFSA suffers from the local search problem, the HC technique is incorporated to it to increase the local searching process. The proposed AFSA-HC algorithm involves four main processes to maximize throughput in the network, namely node initialization, AFSA-HC based node clustering, deflate algorithm based data aggregation, and hybrid data transmission. The incorporation of these processes helps to achieve maximum energy efficiency and throughput. A comprehensive experimental analysis is carried out to verify the effective performance of the presented AFSA-HC technique.

2.2 The Proposed AFSA-HC Technique

The proposed AFSA-HC algorithm involves four main processes to maximize throughput in the network, namely node initialization, AFSA-HC based node clustering, deflate algorithm based data aggregation, and hybrid data transmission. Figure 2.2 shows the working principle of the proposed AFSA-HC algorithm. As shown in the figure, once the nodes are deployed and initialized, the AFSA-HC algorithm gets executed and constructs the clusters. The AFSA-HC algorithm selects the cluster heads (CHs) effectively and constructs the clusters. Next, the CHs execute the deflate algorithm to aggregate the data received from the CMs.

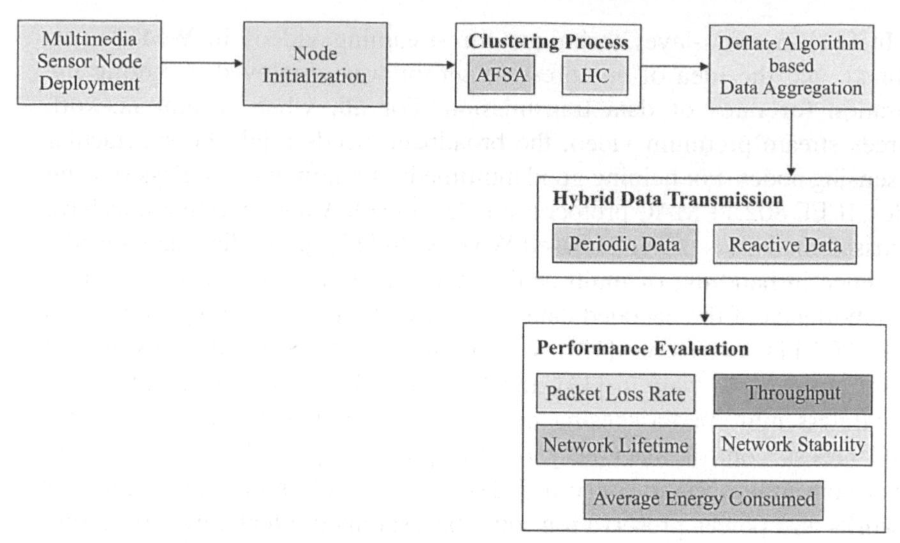

Figure 2.2 Block diagram of AFSA-HC technique.

At last, hybrid data transmission strategy is applied where the data will be transmitted in both proactive and reactive ways.

2.2.1 AFSA-HC Based Clustering Phase

Optimization methods are utilized for minimizing or maximizing the result of a function by changing its aspects. Every suitable value to this issue is known as feasible results, and the better one is the best result. Every swarm intelligence (SI) technique is dependent on population; as their iterative process deals with enhancing the place of separate in population and following, their progress to the optimal positions. AFSA is the most desirable SI technique with benefits such as global explore ability, robustness, tolerant rate of parameter setting, etc., and it can also be verified to be non-sensitive to starting values [16].

The AFSA imitates the performance of fish swarm prey and it is establishing to meet very quickly. A visual concept of AFSA is shown in Figure 2.3. An explore initiates with arbitrary results mapped to all fishes. Fitness is estimated and the fish following method are started and fish following another fish through optimal result. When the result is sub-optimal, swarming methods are started. When the preferred result is not joining, prey procedure is started. This method is continued until the needed threshold or termination conditions are met.

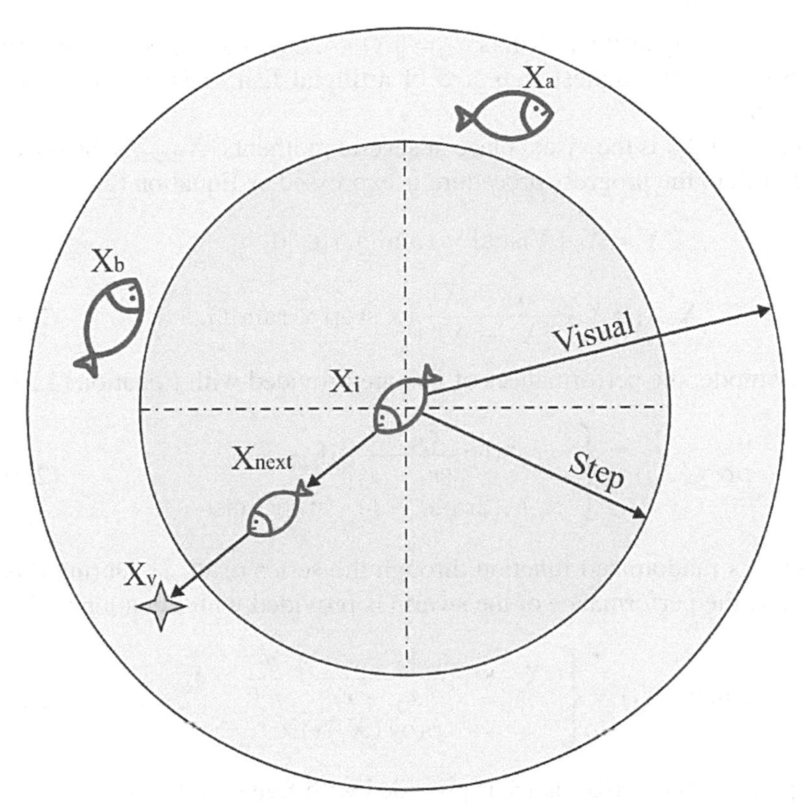

Figure 2.3 Vision concept of the AFSA.

The AFSA has four functions, which have modeled the nature of fish in fish swarm. An initial function is free-move action: as in environment, fish travels generously in swarm if it does not get wounded. The second function is prey action where it utilizes its vision sensing, smelling, and some other accessible sensing devices on the human body. In AFSA, the region where an artificial fish senses the prey is modeled as neighborhoods through a visual-sized radius. The third function is a when a fish determines food; another swarm members and following it to attain the food. The final function is the swarm action that imitates fishes as fish everlastingly attempt to be in the swarm and does not leave to be defended from hunters. The food stability degree in the water region is AFSA objective function.

Assume that the condition vector of AFSA is $(x_1, x_2, \ldots, x_n)$, where $x_1, x_2, \ldots x_n$ is the position of the fish. The heuristic used to represent food focus in this place of fish is $y{=}f\ (x)$, where y is the objective function value.

The distance among artificial fish is $d_{ij}= \|X_i - X_j\|$, i and j is an arbitrary fish. Step means the highest step size of artificial fish. a is the degree of congestion factor.

Assume that X_v is the visual place at several moments. X_{next} is the novel place. After that, the progress procedure is expressed as Equation (2.1):

$$X_v{=}X_i{+}\text{Visual} \times \text{rand}(), i \in [0, n]$$

$$\text{X}_{\text{next}}{=}X{+}\frac{X_v - X}{\|X_v - X\| \,|} \times \text{step} \times \text{rand}(). \tag{2.1}$$

In the prey mode, the performances of fish are provided with Equation (2.2):

$$\text{prey}\,(X_i) = \begin{cases} x_i{+}\text{step}\dfrac{x_j - x_i}{\|x_j - x_i\|}\text{if } y_j - y_i \\ x_i{+}\ (2\text{rand} - 1)\ .\ \text{step} \quad \text{else} \end{cases} . \tag{2.2}$$

Now "rand" is randomized function through the series of $[0, 1]$. During this swarm stage, the performance of the swarm is provided with Equation (2.3):

$$\text{swarm}\,(X_i) = \begin{cases} X_i{+}\text{step}\dfrac{x_j - x_i}{\|x_j - x_i\|}\text{if } \dfrac{y_c}{nf} > \delta y_i \\ \text{prey}\,(X_i)\ \text{else} \end{cases} . \tag{2.3}$$

Following phase, the performance is provided with Equation (2.4):

$$\text{follow}\,(X_i) = \begin{cases} X_i{+}\text{step}\dfrac{x_{\max} - x_i}{\|x_{\max} - x_i\|}\text{if } \dfrac{y\ \max}{nf}{>}\delta y_i \\ \text{prey}\,(X_i)\ \text{else} \end{cases} . \tag{2.4}$$

The three steps mentioned make sure that both global as well as local explore and the explore direction subsequent to the optimal food source are obtained. The flowchart of the AFSA is shown in Figure 2.4. During this work, AFSA is altered as discussed below.

An HC technique was utilized for "local exploration" to achieve the local optimum. As HC techniques are simpler for implementation and provide more flexibility in the action of particles, the AFSA is integrated to HC. The HC techniques utilize arbitrary local search for defining the direction as well as size of every novel phase [17]. The process involved in the HC technique is listed as follows.

- Initialization: in particular, the neighborhood function ϕ and choose an initial solution I_0;

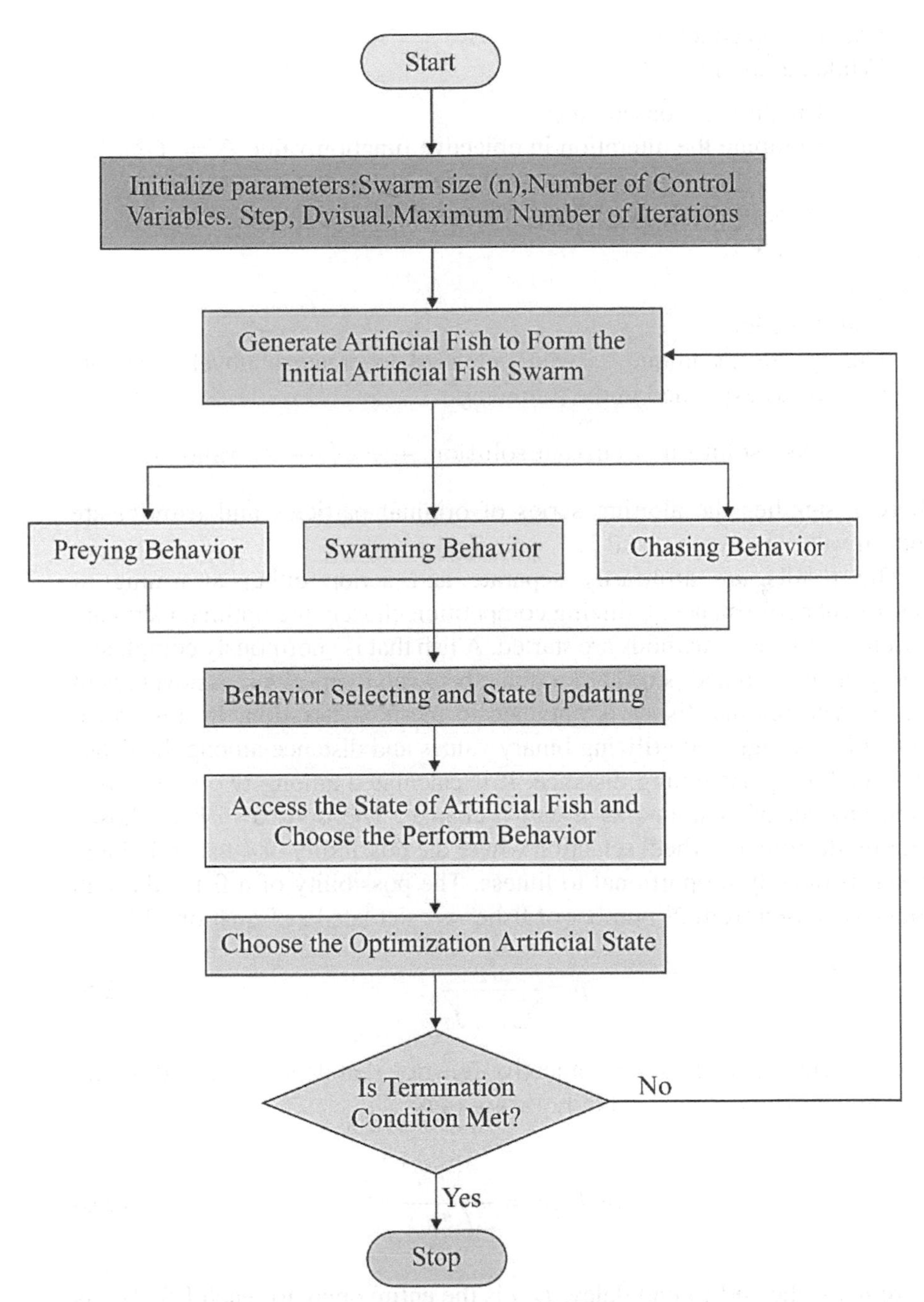

Figure 2.4 Flowchart of AFSA.

- Put the loop counter $z = 1$;
- While iteration $z < Z$;
  - Create $I_z \in \phi$ based on ϕ;
  - Compute the alteration in objective function value $\Delta = C(I_z) - C(I_{z-1})$;
  - Accept solution I_z if $\Delta < O$;
  - $z = z + 1$;
- End
- End Procedure
- During this technique, ϕ were intended to create a novel candidate solution as explained in the following.

$$\text{Newsolution} = \text{current solution} + r * (1 - 2 * \text{rand}\,())$$

where r signifies the altering series of original particles and $\text{Rand}()$ are random values between 0 and 1.

The results are arbitrarily separate and action either swarming or subsequent performance. Utilizing competition chosen, the optimal fishes are chosen and the prey methods are started. A fish that is enormously completely in prey are chosen and permitted to mate between themselves. A novel result created with optimal fishes is obtained to the next iteration. In this paper, the results are signified utilizing binary values and distance among the fishes calculated using Hamming distance. It is calculated among two strings u, v is the amount of positions as u and v change. The optimal fish is chosen utilizing the roulette wheel selection where the possibility of a fish individual chosen is directly proportional to fitness. The possibility of a fish i through fitness f_i chosen from N number of fishes is calculated as Equation (2.5):

$$p_i = \frac{f_i}{\sum_{j=1}^{N} f_j}. \tag{2.5}$$

For enhancing the QoS, a multiobjective function that depends on end-to-end delay and energy is presented in Equation (2.6):

$$\min f\,(\mathrm{x_i}) = \frac{e^{-\left(\frac{D_{td}}{D_m}\right)^2}}{\left(\frac{E_{ri}}{E_i}\right)} \tag{2.6}$$

where d_{ei} is the end-to-end delay, D_{td} is the entire delay to reach BS, D_m is the maximum delay in the network to achieve BS, E_{ri} is the residual energy in the CH, and E_i is the primary energy. Some of the assumptions are listed below:

- Every node is situated arbitrarily in the network.
- All nodes have the similar primary energy.
- Every fish is unisex.
- Because fishes are unisexual in nature, mating is feasible between two fishes.

As the result spaces are signified in a binary format, a transfer function is needed to flip the bit if the fish modifies status. In this paper, a novel transfer function is presented for flipping the bits provided with Equation (2.7):

$$\text{transfer}\,(X_i) = \frac{1}{(1 + \mathrm{e} - \tanh\,(x_i))}. \tag{2.7}$$

For achieving flipping, an arbitrary number is created between 0 and 1; the bits are flipped when the arbitrary number is below the transfer function provided with Equation (2.8):

$$\begin{cases} 1, & \text{if } p\,(0,\ 1) < \text{transfer}\,(x_i) \\ 0, & \text{otherwise} \end{cases}. \tag{2.8}$$

2.2.2 Deflate-Based Data Aggregation Phase

Every CH executes the deflate algorithm to aggregate the data received from its CMs. Deflate is a lossless encoding approach, which make use of LZSS and Huffman coding [18]. A deflate stream comprises a sequence of blocks where each block comes with a three-bit header.

i) First bit: Last-block-in-stream marker: (1: This is the final block in the stream. 0: Several blocks exist for processing subsequent to this one).
ii) Second and third bits: Encoding model is utilized for this block. The saved block option adds lower overhead and is utilized for incompressible data format.

The highly compressible data ends up with the encoding of dynamic Huffman encoding, which generates an optimal Huffman tree customized for every block of data independently. The guideless for the generation of the essential Huffman tree right away tracks the block header. The static Huffman option is utilized for short message, where the fixed saving attained by discarding the tree outweighs the loss of compression loss because of the utilization of a non-optimal code. Generally, the compression is done by the following two steps:

- the matching and replacement of replica strings with pointers;
- replacement of symbols with novel, weighted symbols using the frequency of utilization.

2.2.3 Hybrid Data Transmission Phase

Once the CHs aggregate the data, hybrid data transmission phase begins. Initially, the CHs will broadcast the subsequent variables:

Attribute(A): It is the user-interested parameter to obtain the data.

Threshold values: It comprises hard threshold (HT) and soft threshold (ST). HT is a certain value of a parameter and the value exceeding the HT will be transmitted. The ST indicates a slight variation in the value of the attribute, which triggers the node of transmitting the data again.

Schedule: For scheduling purposes, Time Division Multiple Access (TDMA) is employed where a slot is allocated to every node.

Timer: It defines the maximum duration between two consecutive reports transmitted by a node. It is a multiple of the TDMA schedule length and it is applied for proactive or periodic data transformation.

By the use of hybrid data transmission, the unwanted and redundant transmission of data can be eliminated. It results in maximum energy efficiency and the throughput can be enhanced.

2.3 Performance Validation

A comprehensive experimental analysis is carried out to verify the effective performance of the presented AFSA-HC technique. The simulation outcome inferred that the AFSA-HC technique has attained maximum results in terms of different measures namely energy consumption, throughput, network lifetime, network stability, and packet loss. The parameter settings of the AFSA-HC technique are provided in Table 2.1.

Table 2.2 provides the results analysis of the AFSA-HC model in terms of different measures under varying number of nodes. Figure 2.5 shows the analysis of the AFSA-HC model with existing models in terms of network stability, which refers to the round number at which all the nodes in the

Table 2.1 Parameter settings.

Parameters	Values
Node count	100–1000
Initial Energy	0.50 J
Amplification coefficient	10 pJ/b
Circuit loss	50 nJ/b
Packet size	5000 b
Control packet size	105 b
Fish population	18

Table 2.2 Results analysis of proposed AFSA-HC with state-of-the-art methods.

Nodes	Network stability					Network lifetime				
	LEACH	GA	BFSA	BFSA-TF	AFSA-HC	LEACH	GA	BFSA	BFSA-TF	AFSA-HC
250	220	420	495	520	550	725	810	830	835	860
500	240	450	550	598	620	840	875	865	880	920
750	260	620	620	680	695	875	930	895	935	953
1000	280	695	670	698	720	900	955	972	980	991
Nodes	**Average energy consumed (J)**					**Throughput**				
	LEACH	GA	BFSA	BFSA-TF	AFSA-HC	LEACH	GA	BFSA	BFSA-TF	AFSA-HC
250	0.118	0.058	0.041	0.037	0.025	12,487	15,876	21,658	23,856	27,563
500	0.13	0.061	0.045	0.0392	0.027	15,890	20,540	23,950	26,900	30,850
750	0.14	0.075	0.058	0.047	0.03	25,743	32,650	38,500	40,890	45,950
1000	0.15	0.079	0.059	0.05	0.038	31,170	38,310	54,652	56,990	58,716

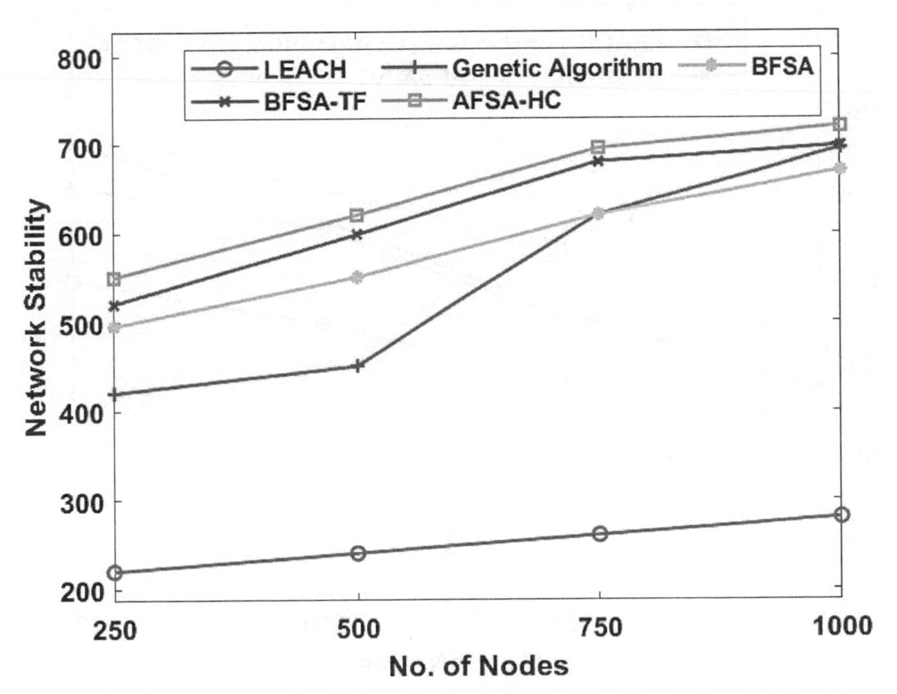

Figure 2.5 Network stability analysis of AFSA-HC with existing models.

network are operative. The network stability should be high for an effective clustering technique. The figure indicated that the AFSA-HC model has showed higher network stability over all the other methods under varying number of nodes. For instance, under the node count of 250, the LEACH protocol has obtained a least network stability of 220 rounds. At the same time, a slightly better network stability of 420 rounds has been attained by the GA, whereas even higher network stability of 495 rounds has been offered by the BFSA technique. Besides, the BFSA-TF has tried to show effective network stability by attaining a maximum of 520 rounds. However, the presented AFSA-HC technique has indicated maximum network stability of 550 rounds. These values ensured the effective performance of the AFSA-HC technique in terms of network stability.

Figure 2.6 demonstrates the analysis of the AFSA-HC method with existing models with respect to network lifetime that refers to the round number at which every node in the network becomes dead. The network lifetime can be high for an effective clustering method. The figure signified that the AFSA-HC model has depicted maximum network lifetime over all

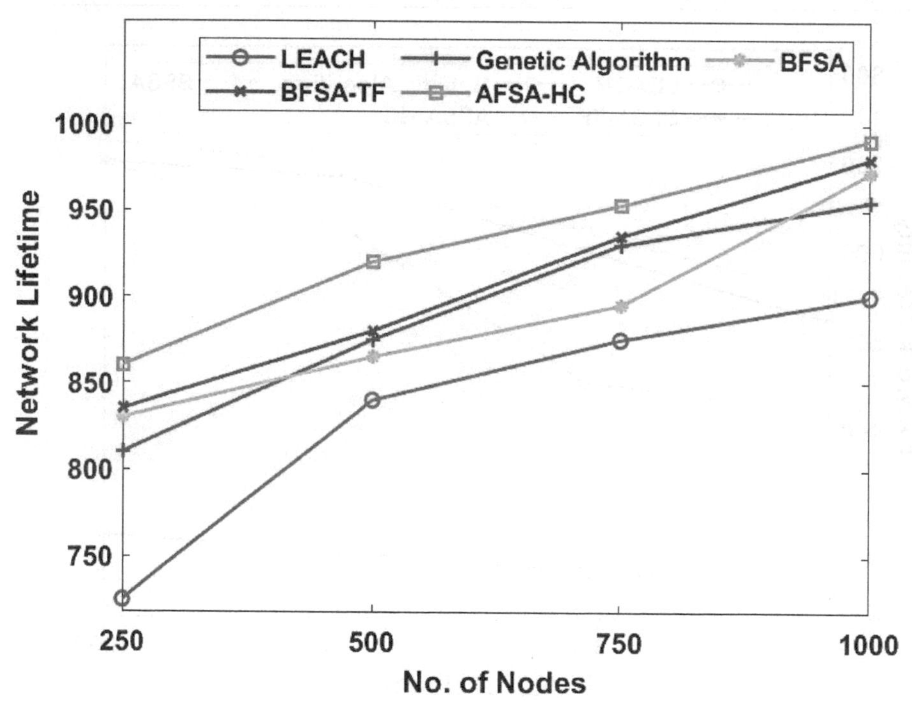

Figure 2.6 Network lifetime analysis of AFSA-HC with existing models.

the other methods under varying number of nodes. For the case under the node count of 250, the LEACH protocol has achieved a small network lifetime of 725 rounds. At the same time, a somewhat better network lifetime of 810 rounds has been obtained by the GA, whereas even maximum network lifetime of 830 rounds has been given by the BFSA technique. Also, the BFSA-TF has tried to illustrate effective network lifetime by obtaining the highest of 835 rounds. But the presented AFSA-HC technique has signified maximum network lifetime of 860 rounds. These values ensure the effective performance of the AFSA-HC technique in terms of network lifetime.

Figure 2.7 depicts the analysis of the AFSA-HC model with predefining models in terms of average energy consumption, and it must be minimum for an effective clustering model. The figure signified that the AFSA-HC model has illustrated less average energy consumption over all the other techniques under different number of nodes. For instance, under the node count of 250, the LEACH protocol has achieved the highest average

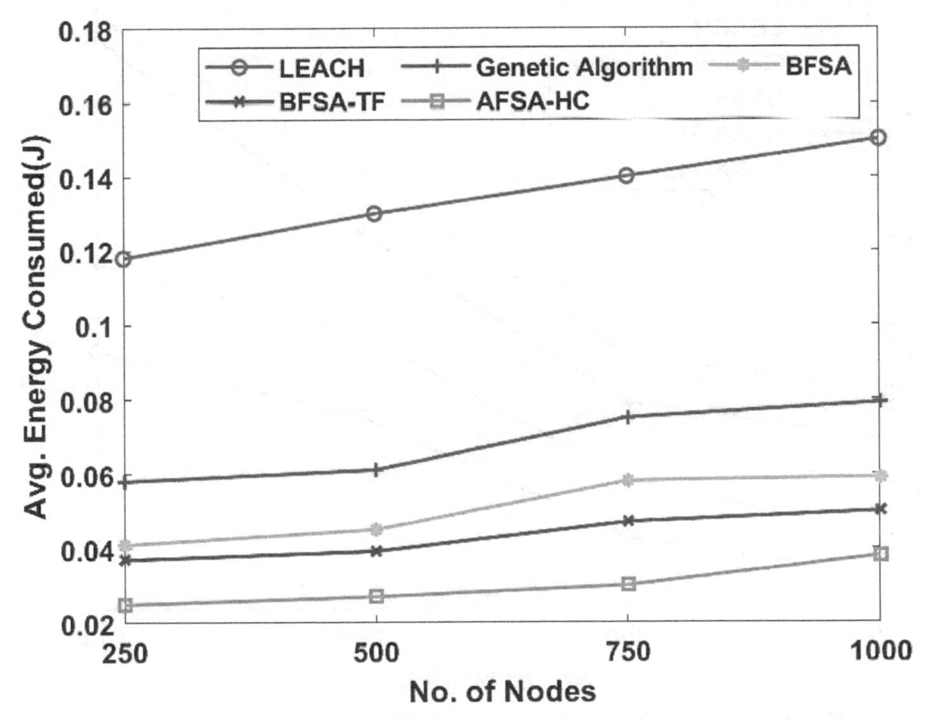

Figure 2.7 Average energy consumed analysis of AFSA-HC with existing models.

energy consumption of 0.118 J. Simultaneously, a slightly lower average energy consumption of 0.058 J has been obtained by the GA, whereas even minimum average energy consumption of 0.041 J has been supplied by the BFSA method. In addition, the BFSA-TF has tried to illustrate effective average energy consumption by obtaining 0.037 J. Conversely, the proposed AFSA-HC method has denoted significantly minimum energy consumed of 0.025 rounds. These values ensured the effective performance of the AFSA-HC technique with respect to the average energy consumption.

Figure 2.8 illustrates the analysis of the AFSA-HC model with existing models with respect to throughput that refers to the entire number of packets that completely achieved the destination. The throughput must be high for an effective clustering technique. The figure signified that the AFSA-HC model has illustrated maximum throughput over all the other techniques

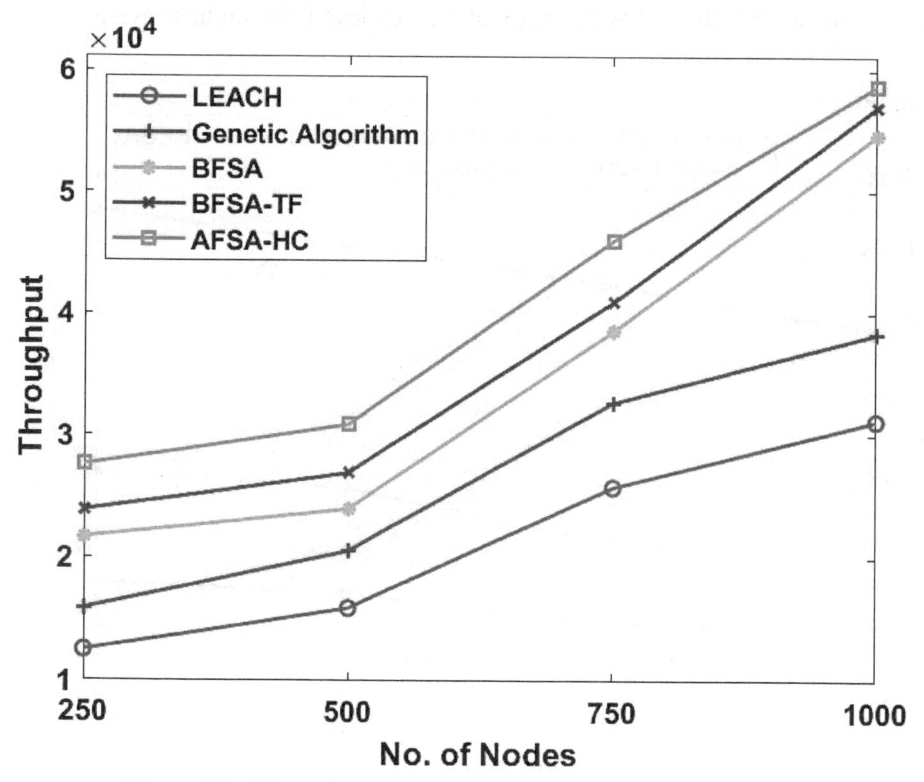

Figure 2.8 Throughput analysis of AFSA-HC with existing models.

under varying number of nodes. For instance, under the node count of 250, the LEACH protocol has attained a minimum throughput of 12,487. Likewise, a somewhat best throughput of 15,876 has been achieved by the GA, whereas even higher throughput of 21,658 has been supplied by the BFSA technique. Also, the BFSA-TF has tried to depict effective throughput by obtaining the highest throughput of 23,856. However, the projected AFSA-HC method has represented the highest throughput of 27,563. These values ensure the effective performance of the AFSA-HC technique in terms of throughput.

Figure 2.9 showcases the analysis of the AFSA-HC model with existing models with respect to packet loss, and it must be low for an effective clustering technique. The figure signified that the AFSA-HC model has depicted less packet loss over all the other techniques under different number of nodes. For instance, under the node count of 250, the BFSA protocol has attained a highest packet loss of 0.96. At the same time, a little minimum packet loss of 0.955 has been achieved by the BFSA-TF, whereas even lesser

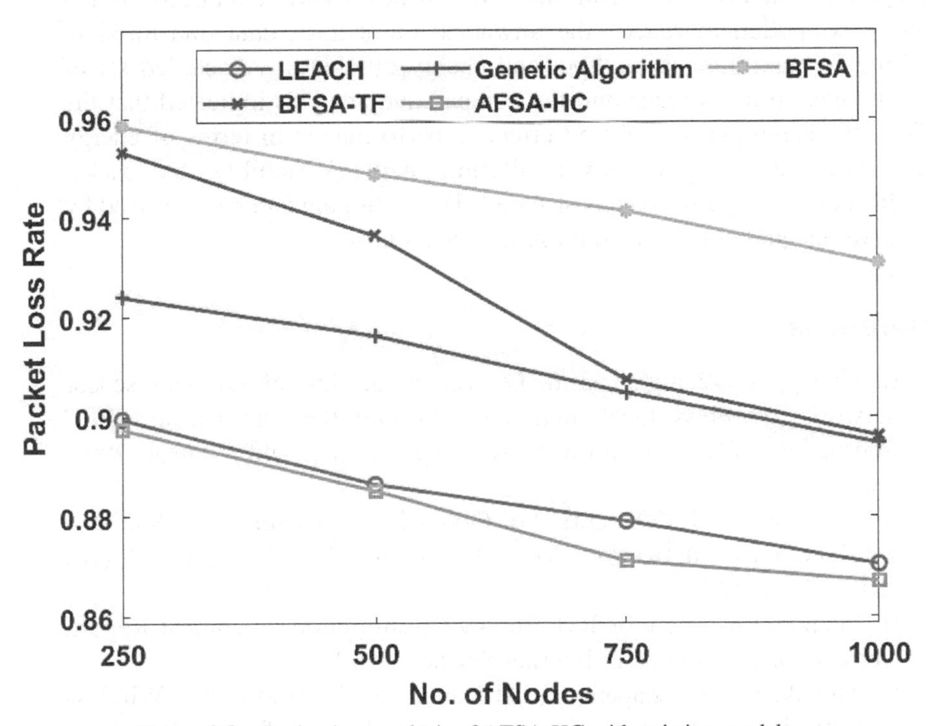

Figure 2.9 Packet loss analysis of AFSA-HC with existing models.

packet loss of 0.923 has been given by the GA method. Also, the LEACH has tried to illustrate effective packet loss by obtaining 0.9. But the proposed AFSA-HC method has demonstrated extensively minimum energy consumed of 0.885. These values ensured the effective performance of the AFSA-HC technique with respect to packet loss.

From the above-mentioned tables and figures, it is ensured that the AFSA-HC model has attained maximum throughput, energy efficiency, network lifetime, and network stability with minimum packet loss rate.

2.4 Conclusion

This paper has presented an effective AFSA-HC based clustering technique for throughput maximization in WMSN. The proposed AFSA-HC algorithm involves four main processes to maximize throughput in the network. Once the nodes are deployed and initialized, the AFSA-HC algorithm gets executed and constructs the clusters. Next, the CHs execute the deflate algorithm to aggregate the data received from the CMs. At last, hybrid data transmission strategy is applied to reduce the unwanted number of data transmission, resulting in maximum throughput and energy efficiency. A detailed set of experimentation takes place and the comparative analysis indicated that the AFSA-HC technique has offered effective performance in terms of energy consumption, throughput, network lifetime, network stability, and packet loss. In future, the performance of AFSA-HC technique can be improved by the use of unequal clustering and routing techniques.

References

[1] L. Cheng, Y. Zhang, T. Lin, Q. Ye, Integration of wireless sensor networks, wireless local area networks and the internet, in: IEEE International Conference on Networking, Sensing and Control, 2004, pp. 462–467.

[2] A. Monsalve, H.L. Vu, Q.B. Vo, Optimal designs for IEEE 802.15.4 wireless sensor networks, Wirel. Commun. Mob. Comput. 13 (18) (2013) 1681–1692.

[3] B. Lannoo, Overview of ICT Energy Consumption, Technical Report, Network of Excellence in Internet Science, 2013.

[4] I. Almalkawi, M. Zapata, J. Al-Karaki, J. Morillo-Pozo, Wireless multimedia sensor networks: current trends and future directions, Sensors 10 (2010) 6662–6717.

[5] Candes Emmanuel J, Wakin Michael B. Introduction to compressive sampling. IEEE Signal Process Mag 2008;25(2):21–30.

[6] Duarte Marco F, Davenport Mark A, Takhar Dharmpal, Laska Jason N, Sun Ting, Kelly Kevin F, et al. Single-pixel imaging via compressive sampling. IEEE Signal Process Mag 2008;25(2):83–91.

[7] A. Boukerche, Y. Du, J. Feng, R. Pazzi, A reliable synchronous transport protocol for wireless image sensor networks, in: IEEE Symposium on Computers and Communications, 2008, pp. 1083–1089.

[8] D.G. Costa, L.A. Guedes, A survey on multimedia-based cross-layer optimization in visual sensor networks, Sensors 11 (5) (2011) 5439–5468.

[9] J.-H. Lee, I.-B. Jung, Adaptive-compression based congestion control technique for wireless sensor networks, Sensors 10 (4) (2010) 2919–2945.

[10] M. Chen, V. Leung, S. Mao, M. Li, Cross-layer and path priority scheduling based real-time video communications over wireless sensor networks, in: IEEE Vehicular Technology Conference, 2008, pp. 2873–2877.

[11] S. Pudlewski, T. Melodia, A. Prasanna, Compressed-sensing-enabled video streaming for wireless multimedia sensor networks, IEEE Trans. Mob. Comput. 11 (6) (2012) 1060–1072.

[12] J. Zhang, J. Ding, Cross-layer optimization for video streaming over wireless multimedia sensor networks, in: International Conference on Computer Application and System Modeling, 2010, pp. 295–298.

[13] C.M.D. Viegas, F. Vasques, P. Portugal, R. Moraes, Real-time communication in IEEE 802.11s mesh networks: simulation assessment considering the interference of non-real-time traffic sources, EURASIP J. Wirel. Commun. Networking 2014 (2014). Article ID 219, 15

[14] H. Touil, Y. Fakhri, M. Benattou, Energy-efficient MAC protocol based on IEEE 802.11e for wireless multimedia sensor networks, in: International Conference on Multimedia Computing and Systems, 2012, pp. 53–58.

[15] N. Chilamkurti, S. Zeadally, R. Soni, G. Giambene, Wireless multimedia delivery over 802.11e with cross-layer optimization techniques, Multim. Tools Appl. 47 (1) (2010) 189–205.

[16] Peng, Z., Dong, K., Yin, H. and Bai, Y., 2018. Modification of Fish Swarm Algorithm Based on Levy Flight and Firefly Behavior. Computational Intelligence and Neuroscience, 2018.

[17] Chen, J., Qin, Z., Liu, Y. and Lu, J., 2005, October. Particle swarm optimization with local search. In *2005 International Conference on Neural Networks and Brain* (Vol. 1, pp. 481-484). IEEE.
[18] Deutsch, P., 1996. Request for comments: 1951. *DEFLATE Compressed Data Format Specification version 1.3. URL= ftp://ftp.uu.net/graphics/png/documents/zlib/zdoc-index. html.*

3

Analysis of Machine Learning Techniques for Spam Detection

Supriya Raheja and Shreya Kasturia

Amity University, Noida
E-mail: supriya.raheja@gmail.com

Abstract

This paper discusses the spam detection and different machine learning models to detect these spam messages. The present work also discusses the testing and evaluation of spam messages. Spam messages have always been very dangerous for computers and networks. They have a very bad effect on the computer security. With the emergence of social media platforms, many people are dependent on emails to communicate, and, with this, there is always a need to detect and prevent spam mails before it enters a user's inbox. The paper also presents the analyses of different machine learning techniques to detect spam messages. Finally, the paper describes the algorithm which is best to detect spam messages.

Spam messages are basically redundant messages which are sent in a large number at once. They can be seen in many forms like free services, cheap SMS plans, lottery, etc. Growing spam messages in your mail can make your inbox filled with ridiculous mails, slow down your Internet speed, and retrieve your private information like credit card details, and has many more drawbacks. Therefore, it is important to prevent it in the best way possible.

Keywords: Spam, ham, Naïve Bayes (NB), random forests (RF), support vector machine (SVM), logistic regression (LR), stemmer, accuracy, confusion matrix.

3.1 Introduction

Emails have become an important aspect in everyone's life with the emergence of various communication social media platforms. Life right now is almost impossible without the use of emails. People would not imagine their work done without the use of emails. Emails have gained huge attention. In 2015, about 205 billion mails were sent and received daily. The number grew to 246 billion in 2019 [1]. Today, more than 85% of the emails are spam where it costs the sender very less amount of time to send, but it costs the recipient or the service provider [26]. Spam mails cost very valuable things like time for humans, servers, and also the valuable ham mails/messages. For every sent mail, there is a quota reserved in the memory space of the server, but the receiver may have less storage, and due to this issue, the receiver may never receive the ham mails because the user is out of storage. The user may also waste their time reading the spam mails. Spam mails can also cause drainage of bandwidth, storage capacity, and productivity which interrupts the ham mails to be delivered to the receiver [7].

With the emergence of new technology come new threats. The word "spam" came into existence in 1937 from "Shoulder Pork HAM," which is a canned precooked meat marketed [2]. Spam emails are reluctant messages sent at once to a large group of users for spamming their inbox to either slow down their Internet speed or retrieve personal details. Therefore, to prevent this type of situation, detecting the spam messages is important. There are many already existing algorithms which make use of machine learning to filter the spam and ham messages [11].

3.1.1 Ham Messages

Ham messages are also seen as everyday messages which the users send to each other as seen in Table 3.1. These messages are not junk and are ought to be received by the user [3], [20].

3.1.2 Spam Messages

Spam messages are messages including words like "free" which are detected by different algorithms with relevance to the other words in the message as seen in Table 3.2.

Table 3.1 Ham messages.

I am free this Sunday, hope to meet you at my house.
Okay I can try to come.
I am doing good, how are you?

Table 3.2 Spam messages

Congratulations, 50% off on successful registration for your new car.
Want a good credit score? Click on the link below to get your score in 5 minutes.
Puma shoes sale, order now! Free on orders above Rs. 2000.

"Are you free on Saturday?" is a ham message while "free melodies and ringtones." is a spam message. Both use the word free but depict different expression [3].

Even though many spam detection techniques/algorithms have been proposed and updated every year, spammers are finding ways to bypass the filter with the use of word obfuscation and statistical poisoning [1].

To detect such spam messages, it is important to train and test the dataset using different algorithms. The algorithm with the least complexity and most efficiency is considered to detect more of spam messages than the rest. Hence, it is important to detect every spam message.

3.2 Types of Spam Attack

There are various spam attacks initiated by the spammers on multiple users daily. They include attacks like phishing, spoofing, clickjacking, whaling, etc. Some of the spammers even try to camouflage the spam message behind an image to bypass the spam detection system. Many companies have also noticed their emails being marked as spam which was not a spam by the anti-spam system which is an expensive false positive detection case [2].

Some types of spam attacks have been further explained.

3.2.1 Email Phishing

It is the most common attack in the recent times. The spammer aims to the C part ("From") or B part of the mail. They aim to show their message such that the user assumes that it is sent by a person they already know [25]. They also interfere with the domain in the HELO statement such that it appears as if it has been sent from a domain that is known, which indicates that spoofing may occur in B part of the data [2].

3.2.2 Spear Phishing

It is the general form of email phishing which tricks the authorized looking messages. These mails may lead the users to open a link and enter their personal or financial details. The spammer aims to the D part ("Body") of the mail. The spammers can even put malicious malwares in user's system just by an attachment. Some of the spammers use social engineering tricks without the use of any links or attachments, which will make the user do some action that will benefit the spammer [25]. Spam phishing is often done by using personal information about the user and sending those mails to what seems to be a reliable source. These attacks are difficult to detect on existing systems and are used to create a dangerous form of attack called "Advanced Permanent Threats" [2].

3.2.3 Whaling

It is also a type of phishing which aims to target the high-level groups of users in a company like CEO of the company or any executive having a high-level clearance in the company. The spammers target the D part ("Body") of the mail and affect the whole company or a user complaint by aiming at the high-level authorities. They will send the mail from a fake origin impersonating as an authorized business establishment luring the company's executive or CEO to do something which will benefit the spammers. It has the same risks as the other phishing attacks [2], [17].

3.3 Spammer Methods

The spammers find the emails of the users by using specialized software. This software collects the addresses of the user from different discussion groups. If not by this method, the spammer buys the address of the user from other spammers [13]. Table 3.3 presents some methods by which the spammers try to spam a user's mail.

3.4 Some Prevention Methods From User End

3.4.1 Protect Email Addresses

Robots and crawlers can easily collect thousands of emails from websites. Even humans try to collect the email addresses when they are in a need of sign-ups. Users can prevent their email address being used in their list by some directives. Users can use "@" instead of "at" and "." instead of "dot." Also, they should more often use image to share their email address to someone.

Table 3.3 Spammer methods.

Methods	Description
Hiding the text	Spammers spam the message by hiding the text behind the JavaScript message, splitting the words, see attached file, etc.
URL hiding	Encoding the URL, copy, and paste.
Zombies and botnets	These are PCs on the Internet which spam using malwares and viruses.
Bayesian sneaking and poisoning	It will look like it is not a spam message since the spammer here uses words which are not spam and "poison" the filter.
Social engineering	They try to convince the user that the mail is not spam by bypassing the spam filter, e.g., faking legitimacy.
IP address	Spammer address borrows an IP address that has good reputation.
Offshore ISPs	This lacks in security measures.
Third-party mailback software	They use mail back application on simple websites.
Falsified header information	They use false heading for their message.
Obfuscation	They use HTML links and innovative symbols to split the messages obscuring the words.
Vertical slicing	Writing the text in messages vertically.
HTML manipulation	They manipulate the HTML format to bypass the spam filter.
HTML encoding	They use encoding measure like Base64 to make binary attachments to text.
JavaScript messages	They write the spam text inside the JavaScript snippet which will pop up once the mail is opened.
ASCII art	Using letter glyphs of the alphabets and sending image to send the spam message.
URL address or redirect URL	They add URL address in the message to bypass the spam filter and spam the user using the website.
Encrypted messages	They encrypt the message which only decrypts when it reaches the inbox of the user.

3.4.2 Preventing Spam from Being Sent

Spammers usually send unsolicited mails to the users using Internet PCs also known as zombies. A user can prevent these mails by checking their computers' security holes and by also blocking the SMTP and proxy relays.

3.4.3 Block Spam to be Delivered

To reject these spam messages, ISPs adopt different methods. One such method is checking the authentication of the sender using "Sender Policy Framework (SPF)" or by blocking the known spammer's sources with the help of IP and Domain Name System (DNS).

3.4.4 Identify and Separate Spam After Delivery

There are mainly two methods which are effective in identifying and separating spam mails.

3.4.4.1 Targeted Link Analysis

Most of the spammers use links instead of text in their mails to spam the inbox of the user which will direct the user to a website which is spam. Analyzing these mails by the users beforehand will prevent such a case to happen.

3.4.4.2 Bayesian Filters

Bayesian filters rely on adaptive filtering algorithms. Users can adapt this algorithm for separating the spam mails from the ham mails [19].

3.4.5 Report Spam

Before deleting the spam mails, the users can report the spam mails so that the spammer will not be able to send any more mails and also they get reported to the ISP and government agencies handing such issues [13].

3.5 Machine Learning Algorithms

Machine learning is a method to make all computational instruments to act on a problem without being explicitly coded. It is a great bonus to tackle from spam due to its ability to adapt with time [14]. As the concepts and operating mechanism are advancing, the already existing spam detection mechanism are rendering useless. This phenomenon is called "Concept Drift." Machine learning algorithms can be of two types – supervised and unsupervised [18], [23]. Following are some algorithms for detecting spam mails.

3.5.1 Naïve Bayes (NB)

It is the most common and easy form of Bayesian method to detect spam. It utilizes probabilistic formula, i.e., Bayes theorem. It observes each feature in dataset individually ignoring the other features in dataset and produces a statistics from each class [4]. This algorithm generally holds well for high-dimensional input dataset. Multinomial NB which is the advanced version of NB provides an additional feature of independence between class and document length [5]. It is a part of supervised learning. According to this algorithm, a class has the highest probability than the posterior probability [6]. The formula evaluates the posterior probability as shown in Equation (3.1) where "spam" denotes the spam mails and "word" denotes the spam word defined by the programmer [16]. The numerator donates the probability of message being spam and containing a spam word, whereas the denominator is the overall probability of an email containing a spam word [15],[24].

$$\frac{P\left(spam \mid word\right) = P\left(word \mid spam\right) P\left(spam\right)}{P\left(word\right)}. \tag{3.1}$$

3.5.2 Random Forests (RF)

It is basically a group of decision tress placed in order to remove the problem of over fitting. In this algorithm, every decision tree will generate its own prediction to the problem which will be different from other trees and then we generalize the final result by averaging the predictions generated by the decision trees [4]. To decide the final prediction and to classify the new object, voting is done and the class with maximum votes decides [5]. It uses bagging to implement the ensemble learning. RF can take into account thousands of data points' features without deletion of any features. It can also calculate the generalized error producing an internal unbiased approximation. For the outliers and noise, RF is very robust. This algorithm has very less complexity as compared to other machine learning algorithms. Moreover, it has less computability as compared to other tree classification methods [21]. The final prediction which is done by voting can be calculated by Equation (3.2) where "$C_{r\,f}^{B}$" denotes the class prediction of *b*th RF [27].

$$C_{r\,f}^{B} = \text{majority vote}\left\{C_B\left(x\right)\right\}_1^B. \tag{3.2}$$

3.5.3 Support Vector Machine (SVM)

This algorithm divides data into groups of related types based on class values using a linear boundary called a hyperplane [3]. It plots data from a dataset in n-dimensional space, assuming each value in the coordinates representing the value of the feature [5]. It is a part of supervised learning. Though it is supervised learning, SVM is also being used in unsupervised clustering. The algorithm may have many hyperplanes, but the perfect one is the one which is the farthest from each class's training dataset. Margins can be estimated using distance from the hyperplane and the nearest data points from both sides [10]. Data points cannot be within margins. The larger the margins are, the better is the model to predict for test data. To calculate the margins, this algorithm makes use of support vectors which are the data points on or closest to the margins. It is very important to optimize the kernel type and kernel parameters to classify emails as spam or ham [2]. SVM decision function is depicted in Equation (3.3). Here, the test dataset's features are denoted by "*y*," support vectors are denoted by "x_i," constant bias by "*n*," weights by "m_i," and the kernel function by "$G(x_i,y)$" which plots the dataset in an n-dimensional feature space [28].

$$F(y) = \sum_{i=1}^{N} mG(x_i, y) + n. \tag{3.3}$$

3.5.4 Logistic Regression (LR)

It is the most favored algorithm involving binary results and is a linear algorithm. An independent set of variables is used to calculate the discrete values. In LR, the output, or event probability, is predicted to fit along a logistic function. The most used LR is sigmoid [5]. LR, rather than choosing the variables which reduce the sum of squared errors, chooses variables which increase the scope of evaluating the dataset [4]. LR gives the best accuracy (98%) when the dataset has data points less than a thousand. To maintain the consistency of the performance, "multiple instance logistic regression (MILR)," which is another version of LR, must be used as it serves an accuracy of 93.3%–94.6% up to 2500 data points [2]. LR can take as input any real-valued number and plot it into an S-shaped curve having a value between 0 and 1 but never exactly on 0 or 1 [22]. The best fitting model can be described by Equation (3.4), where "*P*" denotes the predicted output, "m_0" is the bias, and "m_1" is single input value's (*x*) coefficient. There is a coefficient "*m*" associated with every column of the data input that must be examined from the training data [7].

$$P = \frac{e^{(m_0 + m_1 * x)}}{(1 + e^{(m_0 + m_1 * x)})}. \tag{3.4}$$

3.6 Methodology

3.6.1 Database Used

"Spam" file is a dataset of 5572 mails which are already classified as spam and ham. The dataset is collected from website [12]. The dataset is formed by "comma-separated values." These files contain one line of text per entry; each line has two columns, i.e., the first column tells us the label of the message (spam or ham) and the second column tells us the text in the message.

3.6.2 Work Flow

Our aim is to analyze the algorithms and train the model with the already existing dataset. First, we load the data from our system in .csv file using pandas read file function. After loading the data, we remove the unnecessary data that include the punctuations as mentioned in Figure 3.1, white spaces,

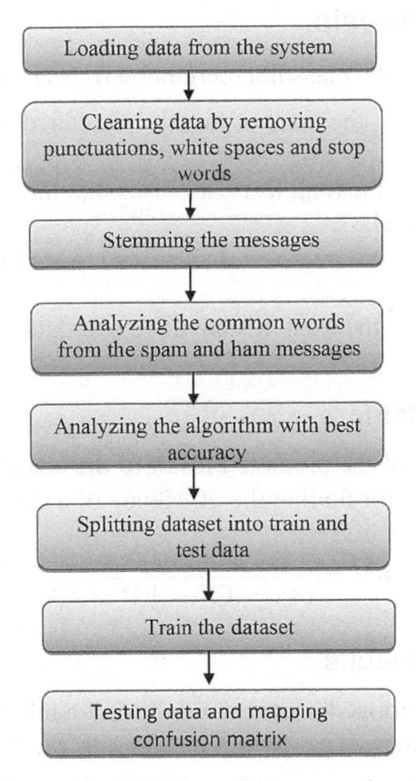

Figure 3.1 Work flow of the proposed work.

and stop words like "the," "an," etc. After this, it is important to stem the messages, i.e., reducing inflection in words to their root form. It will not give any advantage but will not do any harm as well. After the messages are ready to be tested upon, analyzing is important. First, we analyze the most common words in spam and ham messages. Then we analyze the algorithms by checking the accuracy of each one to predict the spam mails or ham mails. After we find the algorithm with best accuracy, we implement it. Here, we divided the dataset into train and test data. The model will take the "train data" and study the features or words which are classifying a mail as spam or ham and we test the data with the "test data" and let the model predict the classification of the mail, i.e., spam or ham. After the model has predicted, we will map a confusion matrix that will tell us the correctness of our model by calculating the right and wrong predictions. These metrics determine the accuracy score of the model [8].

3.7 Results and Analysis

To determine accuracy of the classifier, certain metrics are used:

a. True positive (TP) – the total correctly classified test cases in the prediction.
b. True negative (TN) – the total test data cases that the classifier correctly rejected by the main class.
c. False positive (FP) – the total test data cases rejected by the main class by the classifier.
d. False negative (FN) – the total misclassified test data cases.

3.7.1 Performance Metric

Accuracy – the measure of the predicted result to the actual result. Equation (3.5) denotes the formula for finding the efficiency of a classifier.

$$\text{Accuracy} = \frac{\text{TP} + \text{TN}}{\text{TP} = \text{TN} = \text{FP} = \text{FN}}. \tag{3.5}$$

3.7.2 Experimental Results

By using the above-mentioned metrics, we have found the best classifier having the highest accuracy, i.e., NB. After finding the best classifier, we

	label	text
0	ham	Go until jurong point, crazy.. Available only ...
1	ham	Ok lar... Joking wif u oni...
2	spam	Free entry in 2 a wkly comp to win FA Cup fina...
3	ham	U dun say so early hor... U c already then say...
4	ham	Nah I don't think he goes to usf, he lives aro...

Figure 3.2 Few entries of dataset used.

	text			
	count	unique	top	freq
label				
ham	4825	4516	Sorry, I'll call later	30
spam	747	653	Please call our customer service representativ...	4

Figure 3.3 Grouping by ham and spam mails.

will compute the confusion matrix which helps us determine the correctness of our model by telling the number of TP, TN, FP, and FN. After choosing our algorithm, we have implemented NB and computed the confusion matrix for it [9]. After computing, we also see that the accuracy of the model comes out to be the same if we put the parameters in Equation (3.5).

As illustrated in Figure 3.2, the first five entries from the dataset have been shown. The first column shows the serial number, the second column shows the label, i.e., spam or ham, and the third column shows the messages.

Figure 3.3 describes the grouping of the mails on the basis of their label. It also shows the message count, number of unique mails, top common message, and the frequency of the common message.

The x-axis tells us the label, whereas the y-axis tells us the value count. We can see in Figure 3.4 that the score of ham mails lies between 4000 and 5000 and score of spam mails lies between 0 and 1000.

Figure 3.5 shows us the pie chart of the mails. Here, we can see the percentage of spam and ham mails in the total dataset. If the total mails are 100%, ham mails take 87% and spam mails take 13%.

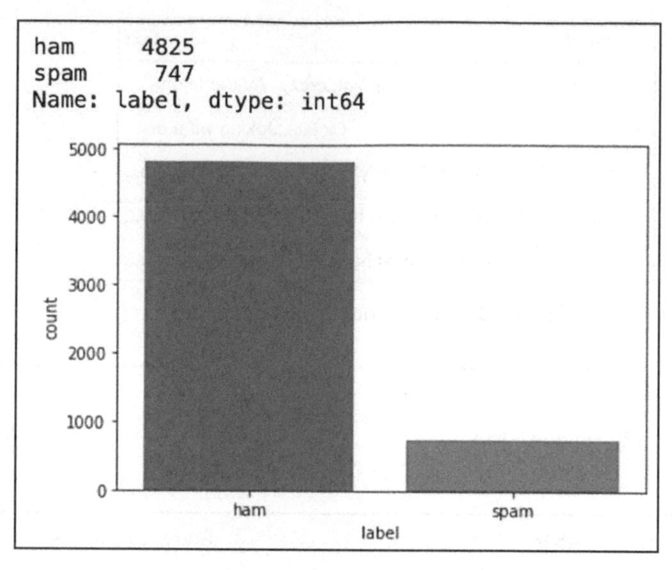

Figure 3.4 Detection of spam and ham messages.

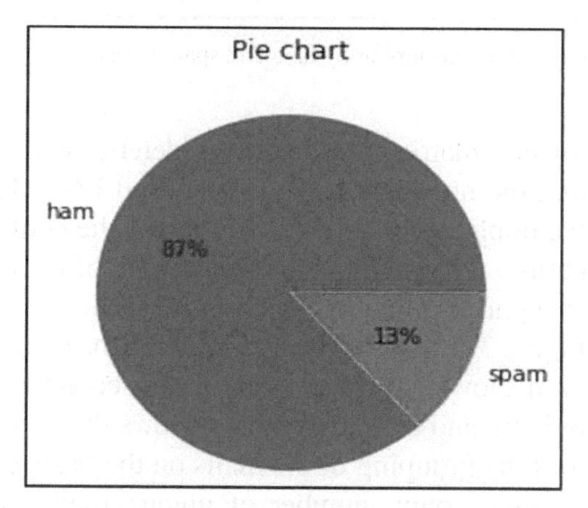

Figure 3.5 Percentage wise distribution.

3.7.2.1 Cleaning Data by Removing Punctuations, White Spaces, and Stop Words

It is important to remove the unnecessary things in a message, which does not affect the classifying of the mails such as punctuations, white spaces, and stop words, i.e., “the,” “he,” “have,” etc.

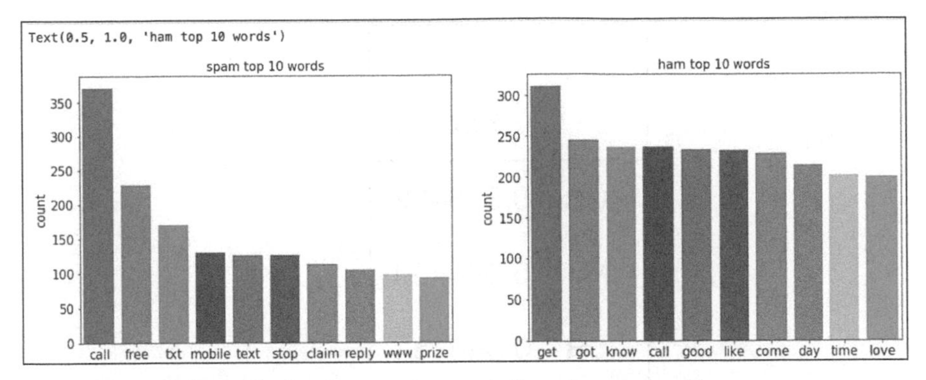

Figure 3.6 Analyzing the common words from the spam and ham messages.

3.7.2.2 Stemming the Messages

Stemming is reducing the inflection in words by stemming them to their root. For example, "playing," "plays," and "played" will be stemmed to "play." It does not give any advantage, but it does not harm either.

3.7.2.3 Analyzing the Common Words from the Spam and Ham Messages

After filtering the messages, we now need to analyze the common words from the ham and spam messages separately. As shown in Figure 3.6, we plot a bar graph where the *x*-axis shows the top 10 spam words and the top 10 ham words from the dataset. The *y*-axis shows the value count of the words.

3.7.3 Analyses of Machine Learning Algorithms

3.7.3.1 Accuracy Score Before Stemming

After analyzing the common words, we need to find out the accuracy of our models to decide which algorithm is the most efficient in detecting spam mails. Figure 3.7 shows the score of each model and we found out that NB model shows the most accuracy [29].

3.7.3.2 Accuracy Score After Stemming

Stemming the messages does not have any advantage nor does it harm anything. It would only give a more precise accuracy score. After stemming, we obtained the score in column "Score2" as shown in Figure 3.8.

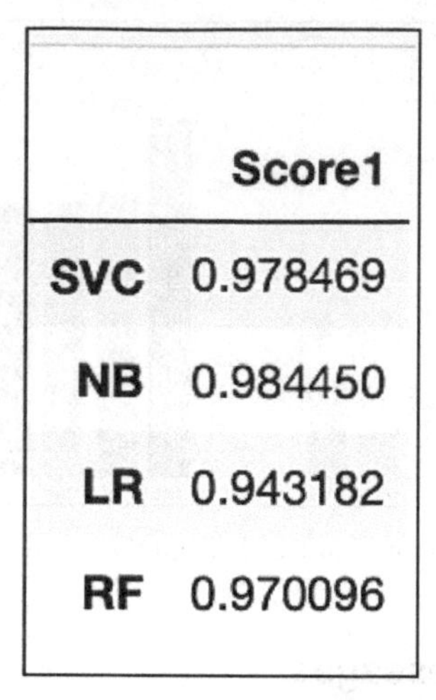

	Score1
SVC	0.978469
NB	0.984450
LR	0.943182
RF	0.970096

Figure 3.7 Accuracy score before stemming.

	Score1	Score2
SVC	0.978469	0.978469
NB	0.984450	0.985048
LR	0.943182	0.946770
RF	0.970096	0.974880

Figure 3.8 Accuracy score after stemming.

Figure 3.9 shows us the plot between the different models and their respective accuracy score. Here, we can clearly see that the NB model has the highest score out of all the other algorithms.

3.7.3.3 Splitting Dataset into Train and Test Data

To train and test our model, we first need to split the dataset into train and test data, i.e., some of the entries will be considered as train data and the rest as test data [30].

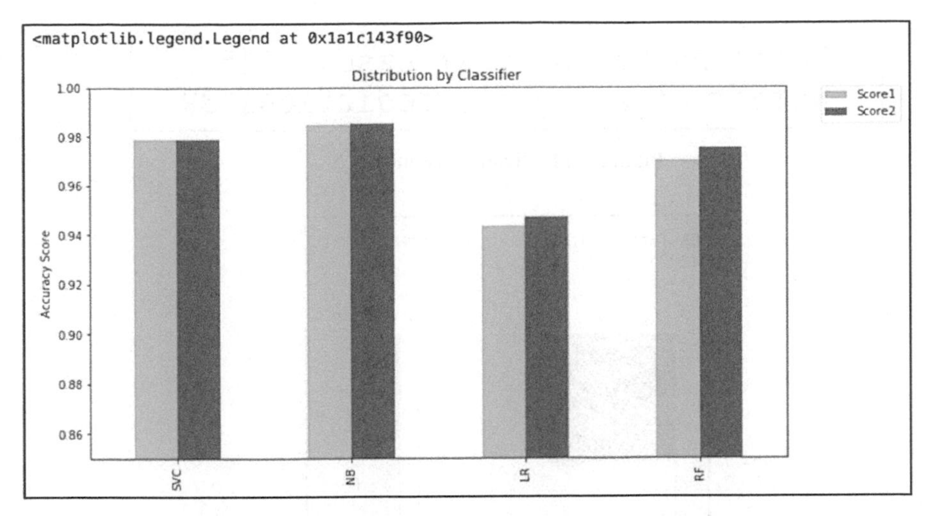

Figure 3.9 Accuracy score using different techniques.

```
Pipeline(memory=None,
         steps=[('vectorizer',
                 TfidfVectorizer(analyzer=<function process at 0x1a1d36d200>,
                                 binary=False, decode_error='strict',
                                 dtype=<class 'numpy.float64'>,
                                 encoding='utf-8', input='content',
                                 lowercase=True, max_df=1.0, max_features=None,
                                 min_df=1, ngram_range=(1, 1), norm='l2',
                                 preprocessor=None, smooth_idf=True,
                                 stop_words=None, strip_accents=None,
                                 sublinear_tf=False,
                                 token_pattern='(?u)\\b\\w\\w+\\b',
                                 tokenizer=None, use_idf=True,
                                 vocabulary=None)),
                ('classifier',
                 MultinomialNB(alpha=1.0, class_prior=None, fit_prior=True))],
         verbose=False)
```

Figure 3.10 Training the model.

We now have to train our model with the train data. Here, pipeline will filter the mails, i.e., vectorize and classify them, and then we train the model on the results obtained as illustrated in Figure 3.10.

Testing of data is important. When we tested the test data after training our model on the train data, we found out the count of total test cases and the test cases that were wrongly classified by our model as shown in Figure 3.11.

```
Total number of test cases 1115
Number of wrong of predictions 39
```

Figure 3.11 Testing the model.

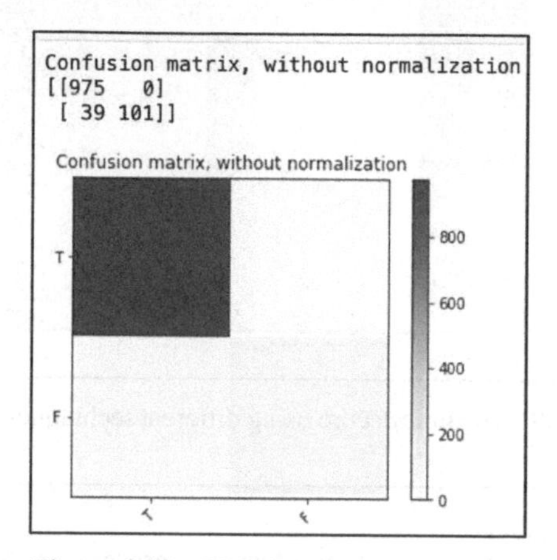

Figure 3.12 Mapping confusion matrix.

3.7.3.4 Mapping Confusion Matrix

To find out more accurate predictions of our model, here, we plot a confusion matrix which will tell us the correctness of our model by telling the number of right and wrong predictions as shown in Figure 3.12.

3.7.3.5 Accuracy

Figure 3.13 has been summarized learning the confusion matrix. The last step is to calculate the accuracy of our model based on the predictions it has made based on the train and test data. We got an accuracy of 96.5%.

```
True Negatives:  975
False Positives:  0
False Negatives:  39
True Positives:  101
```

Figure 3.13 Determining TP, TN, FP, and FN.

3.8 Conclusion and Future Work

This paper discussed the spam messages and different machine learning algorithms in detail. The work also discussed how the spam and ham messages are detected from a set of messages. The work is implemented using python. The model learns the features of the data we provided and then detects if a message is spam or ham, seeing words that are common in the mails. Different machine learning algorithms are evaluated. Results concluded that the NB classifier gives the best accuracy.

By analyzing the program, we have found that NB and SVM give good results for the detection of spam in datasets. Significant results have been obtained. This implementation can further be used in real world for spam detection. Training the model with multiple number of dataset can increase the chance of detecting a spam in a mail. Also, by implementing multiple classifiers, i.e., hybrid, we can increase the efficiency.

References

[1] Peng, W., Huang, L., Jia, J., & Ingram, E. (2018, August). Enhancing the naive bayes spam filter through intelligent text modification detection. In 2018 17th IEEE International Conference On Trust, Security And Privacy In Computing And Communications/12th IEEE International Conference On Big Data Science And Engineering (TrustCom/BigDataSE) (pp. 849-854). IEEE.

[2] Karim, A., Azam, S., Shanmugam, B., Kannoorpatti, K., & Alazab, M. (2019).A Comprehensive Survey for Intelligent Spam Email Detection. IEEE Access, 7, 168261-168295.

[3] Navaney, P., Dubey, G., & Rana, A. (2018, January). SMS Spam Filtering Using Supervised Machine Learning Algorithms. In 2018 8th International Conference on Cloud Computing, Data Science & Engineering (Confluence) (pp. 43-48).IEEE.

[4] Sethi, P., Bhandari, V., & Kohli, B. (2017, October). SMS spam detection and comparison of various machine learning algorithms. In 2017 International Conference on Computing and Communication Technologies for Smart Nation (IC3TSN) (pp. 28-31). IEEE.

[5] Gupta, M., Bakliwal, A., Agarwal, S., & Mehndiratta, P. (2018, August). A Comparative Study of Spam SMS Detection Using Machine Learning Classifiers. In 2018 Eleventh International Conference on Contemporary Computing (IC3) (pp. 1-7). IEEE.

[6] Agarwal, K., & Kumar, T. (2018, June). Email Spam Detection Using Integrated Approach of Naïve Bayes and Particle Swarm Optimization. In 2018 Second International Conference on Intelligent Computing and Control Systems (ICICCS) (pp. 685-690). IEEE.

[7] Bassiouni, M., Ali, M., & El-Dahshan, E. A. (2018). Ham and spam e-mails classification using machine learning techniques. Journal of Applied Security Research, 13(3), 315-331.

[8] Shahariar, G. M., Biswas, S., Omar, F., Shah, F. M., & Hassan, S. B. (2019, October). Spam Review Detection Using Deep Learning. In 2019 IEEE 10th Annual Information Technology, Electronics and Mobile Communication Conference (IEMCON) (pp. 0027-0033).IEEE.

[9] Makkar, A., Garg, S., Kumar, N., Hossain, M. S., Ghoneim, A., & Alrashoud, M. (2020). An Efficient Spam Detection Technique for IoT Devices using Machine Learning. IEEE Transactions on Industrial Informatics.

[10] Olatunji, S. O. (2019). Improved email spam detection model based on support vector machines. Neural Computing and Applications, 31(3), 691-699.

[11] Radovanović, D., & Krstajić, B. (2018, February). Review spam detection using machine learning. In 2018 23rd International Scientific-Professional Conference on Information Technology (IT) (pp. 1-4).IEEE.

[12] https://www.kaggle.com/uciml/sms-spam-collection-dataset/home

[13] Varghese, L., & Jacob, K. P. (2019). An Integrated approach to spam filtering and incremental updation of spam corpus using data mining techniques with modified spell correction algorithm (Doctoral dissertation).

[14] Visani, C., Jadeja, N., & Modi, M. (2017, August). A study on different machine learning techniques for spam review detection. In 2017 International Conference on Energy, Communication, Data Analytics and Soft Computing (ICECDS) (pp. 676-679).IEEE.

[15] Trivedi, S. K. (2016, September). A study of machine learning classifiers for spam detection.In 2016 4th international symposium on computational and business intelligence (ISCBI) (pp. 176-180).IEEE.

[16] Jatana, N., & Sharma, K. (2014, March). Bayesian spam classification: Time efficient radix encoded fragmented database approach. In 2014 International Conference on Computing for Sustainable Global Development (INDIACom) (pp. 939-942).IEEE.

[17] Wang, L., Zeng, G., & Huang, B. (2020). Naive Bayesian Algorithm for Spam Classification Based on Random Forest Method. In Journal of Physics: Conference Series (Vol. 1486, p. 032021).

[18] Ghosh, A., & Senthilrajan, A. (2020). Implementing Naive Bayes Algorithm for Detecting Spam Emails on Datasets.

[19] Ning, B., Junwei, W., & Feng, H. (2019). Spam message classification based on the naïve Bayes classification algorithm. IAENG International Journal of Computer Science, 46(1), 46-53.

[20] Alurkar, A. A., Ranade, S. B., Joshi, S. V., Ranade, S. S., Shinde, G. R., Sonewar, P. A., & Mahalle, P. N. (2019). A Comparative Analysis and Discussion of Email Spam Classification Methods Using Machine Learning Techniques.In Applied Machine Learning for Smart Data Analysis (pp. 185-206).CRC Press.

[21] Goswami, V., Malviya, V., & Sharma, P. (2019, October). Detecting Spam Emails/SMS Using Naive Bayes, Support Vector Machine and Random Forest.In International Conference on Innovative Data Communication Technologies and Application (pp. 608-615).Springer, Cham.

[22] Vinitha, V. S., & Renuka, D. K. (2019, April). Performance Analysis of E-Mail Spam Classification using different Machine Learning Techniques. In 2019 International Conference on Advances in Computing and Communication Engineering (ICACCE) (pp. 1-5). IEEE.

[23] Jáñez-Martino, F., Fidalgo, E., González-Martínez, S., & Velasco-Mata, J. (2020). Classification of Spam Emails through Hierarchical Clustering and Supervised Learning. arXiv preprint arXiv:2005.08773.

[24] Mohammed, Z., JA, M. F., MP, M. I., Basthikodi, M., & Faizabadi, A. R. (2019). A Comparative Study for Spam Classifications in Email Using Naïve Bayes and SVM Algorithm.

[25] Abu-Nimeh, S., Nappa, D., Wang, X., & Nair, S. (2007, October). A comparison of machine learning techniques for phishing detection. In Proceedings of the anti-phishing working groups 2nd annual eCrime researchers summit (pp. 60-69).

[26] Tretyakov, K. (2004, May). Machine learning techniques in spam filtering.In Data Mining Problem-oriented Seminar, MTAT (Vol. 3, No. 177, pp. 60-79).Citeseer.

[27] Alurkar, A. A., Ranade, S. B., Joshi, S. V., Ranade, S. S., Sonewar, P. A., Mahalle, P. N., & Deshpande, A. V. (2017, November). A proposed data science approach for email spam classification using machine learning

techniques. In 2017 Internet of Things Business Models, Users, and Networks (pp. 1-5).IEEE.
[28] Lakshmi, R. D., & Radha, N. (2010). Spam classification using supervised learning techniques. In Proceedings of the 1st Amrita ACM-W Celebration on Women in Computing in India (pp. 1-4).
[29] Saad, O., Darwish, A., & Faraj, R. (2012). A survey of machine learning techniques for Spam filtering.International Journal of Computer Science and Network Security (IJCSNS), 12(2), 66.
[30] Lai, C. C., & Tsai, M. C. (2004, December). An empirical performance comparison of machine learning methods for spam e-mail categorization. In Fourth International Conference on Hybrid Intellig

4

Smart Sensor Based Prognostication of Cardiac Disease Prediction Using Machine Learning Techniques

V. Vijayaganth and M. Naveenkumar

Assistant Professor, Department of Computer Science and Engineering, KPR Institute of Engineering and Technology, Coimbatore, India.
E-mail: kv.vijayaganth@gmail.com, naveenkumar2may94@gmail.com

Abstract

Healthcare sectors are more predominantly growing in the modern decade. People are concerned about their health status because of unhealthy foods and imbalanced diet. People are switching to gymnasium, yoga, and other required healthy life activities. Despite the fact that many healthcare companies collect large quantities of data that contain some hidden information which are used for decision making. In this work we concentrate on predicting cardiovascular disease which is a leading cause of global death. The death rate is low in lower income counties than high income countries. It is mainly due to people's lifestyle in the modern era changing factors such as obesity, diabetes, and others. Early detection of heart disease in the healthcare sector is difficult because it depends on clinical and pathological data. The purpose of this work is to use machine learning (ML) algorithms to implement a solution for earlier prediction of cardiovascular disease. The work comprises various sensors to get the user's medical parameter like resting blood pressure (RBS), serum cholesterol (chol), fasting blood sugar (FBS), max heart rate, etc., and the data are processed with various ML algorithms like random forest, *k*-nearest neighbor (*k*-NN), decision tree (DT), and extreme gradient boost (XGBoost) method. Through these various methods, we have analyzed and compared the accuracy of the disease prediction.

Keywords: Sensors, heart disease, resting blood pressure (RBS), extreme gradient boost (XGBoost).

4.1 Introduction

Health has become one of the difficulties confronting public around the globe. World Health Organization (WHO) stated that the fundamental right is for an individual's proper health. People should take healthy, diet-fit food in the current era [1]. So, proper healthcare services should be provided to keep people fit and healthy. 31% of all deaths globally are heart-related diseases [2]. Diagnosis and treatment of cardiovascular disease are very complex in developing nations due to the absence of diagnostic devices as well as a shortage of doctors and many other resources that affect adequate prediction and medication of heart patients. Recently, computer technology and machine learning (ML) methods exist with this concern to improve the system to support doctors in the preliminary stage in making decisions about heart disease.

Early stage detection of the disease and predicting a person's likelihood of being at risk of heart disease can reduce the rate of death. Health diagnosis is a significant task which must be performed thriving and accurately. Technology must help doctors diagnose better since human consciousness can make false presumptions and unexpected outcomes. Not only does this automation help doctors, but it also warns patients about their fleshly health and to go for a health check-up. Analysis of a disease shows an important role in the healthcare sector. A hospital has a large database that includes the specifics of the patient and remains unexplored with a lot of secret information and expertise. An ML methodology can be used to forecast a disease from the gathered data. It is generally well-defined as investigating the hidden data contained in the dataset, and the information acquired must be unique and useful for others [3]. The healthcare data mining techniques are used to derive useful trends and information in medical data. Medical data is redundant, multi-attributive, incomplete, and is closely related to time.

The problem of making effective use of the huge amount of data becomes a major public health sector problem. Different approaches have been introduced in data mining to discover the harshness of human cardiac disease. Disease frequency is graded using various approaches such as *k*-nearest neighbor (*k*-NN), Naive Bayes (NB), DT, etc. [4], [5], [6]. We have already seen the use of decision trees (DT) to estimate the frequency of heart disease related events [7]. Different methods have been employed to abstract knowledge by using known statistical methods to predict heart disease.

4.2 Literature Survey

Takuji Suzuki *et al.* proposed a wearable wireless sensor to gather healthcare data. This work monitors body surface temperature, pulse, programmable system on a chip (P-SoC), electrocardiogram (ECG), and gesture sensors. These are combined with great performance arm CPU and double mode Bluetooth chips in a slight and lightweight frame. It measures health status and collects data through a network. Simple prototype is attached to a male's chest; ECG and pulse are measured and send over a Bluetooth device [8].

Purnendu Shekhar Pandy proposed a system to detect and predict the stress of humans using ML techniques using Internet of Things (IoT). The prototype detects the stress based on the heart rate. It records the data when the device is connected to the tip of the finger, which reduces human effort in collecting data and privacy and security is maintained [9].

Purushottam *et al.* developed a system to predict heart disease using DTs. The system provides efficient rules to predict disease based on users' requirements. The authors have used Cleveland clinic dataset; it contains 76 attributes, but only 14 attributes are used for prediction [10].

M. Chen *et al.* proposed a disease extrapolation system using ML techniques in big data. Structured and unstructured data are used to model the disease prediction using conventional neural networks. Structured data contains information like age, gender, and life style, and unstructured data contains patient's history like illness, records, and diagnosis [11].

Nam T. Nguyen *et al.* proposed an emotion-based heart disease exploration using heart rate signal. Data is collected from sensors using android application. Naveenkumar M and Vishnu Kumar Kaliappan proposed a speech emotion detection and recognized using mel-frequency using cepstral coefficient. The authors have found that an age from 18 to 30 does not have any heart disease. They discussed the emotions of fear, anger, sadness, disgust, neutral, and happiness [12] [13].

Aakash Chauhan *et al.* used evolutionary learning to study rules for predicting the danger of coronary malady in patients. A comprehensive examination of data mining has remained and carried out to attain better accuracy. Computational intelligence (CI) methods are used to catch the connection among the patients and their diseases. The dataset contains 13 attributes of C level used for this research [14]. Data cleaning is processed to provide quality data using data pre-processing methods in data mining. A precision of 0.53 has been accomplished on a test dataset. Visit design development affiliation mining is applied on a dataset to acquire better affiliation rules. The examination reasoned that more the standards are, the better the forecast of coronary infection.

Chen *et.al.* have built up a framework utilizing an artificial neural network (ANN) that assesses the people's clinical information that help medicinal service experts in exploring the heart-related disease and its threat factors. The authors used the University of California Irvine (UCI) ML repository dataset for prediction. This approach uses three steps. Initially, 13 features are selected from 14 features which include sex, trestbps, chest pain type, blood sugar, thal, resting ECG, exercise-induced angina, slope, and others [15]. Next, classification-based algorithms were applied to the features extracted in the first step. This uses an ANN and achieves 80% accuracy. Finally, to provide user-friendly access, graphical user interface was developed.

Ankita Dewan and Meghna Sharma created an algorithm for prediction using back-propagation which is efficient and hybrid in nature. They proposed a classification technique for non-linear data using neural networks. The proposed algorithm uses back-propagation ANNs with updating technique of backward error propagation [16]. A flaw was identified at implementing optimization algorithms with local minima solutions to improve the accuracy. Heart disease was done with 100% accuracy or with least error using hybrid technique.

Aditi Gavhane *et al.* predicted heart diseases using multilayer perceptron (MLP) that a user's probability of getting coronary artery disease (CAD) with high accuracy [17]. The authors used Cleveland dataset available in UCI ML repository and established using PyCharm IDE. Better accuracy was obtained using an MLP with graphical user interface.

Shashikant U. Ghumbre and Ashok A. Ghatol compared algorithms like support vector machine (SVM), MLP, radial basis function (RBF), and others [18]. The authors proposed SVM and RBF network arrangement to make decisions on heart disease. The authors suggested SVM using sequential minimization optimization which provides better results as compared to ANN. The precision was 85.51% with fivefold cross approval and 85.05% with ten times cross approval.

R. Kavitha and E. Kannan proposed a framework for feature extraction using principal component analysis (PCA) and outlier detection. This method uses wrapper filter for feature subset selection and to improve the performance scoring functions like Pearson's correlation coefficient and Euclidean distance [19]. This framework diagnoses heart disease and reduces the attributes for prediction.

M. Akhil Jabbar *et al.* performed their analysis of heart disease using *k*-NN and genetic algorithm classification. The authors combined both the techniques with six medical and one non-medical dataset. The authors found 60% better accuracy in genetic algorithms than *k*-NN [20]. Big data can be implemented to acquire huge volumes of data and privacy can be maintained [21].

4.3 Proposed Method

Smart sensor based prognostication cardiac disease prediction system (SPCDPS) framework is based on two functions. They fetch data from sensors and apply the knowledge to the data. SPCDPS comprises an "*N*" number of sensor gadgets to bring the information from the human body. Once the information is fetched by the sensor, the data is transferred to the dataset via Arduino through Wi-Fi module. The data from the dataset is pre-processed through various pre-processing methods and the human body state is predicted. After pre-processing the data, SPCDPS splits the information into preparing and testing in the proportion of 80 and 20 models. SPCDPS is provided with various algorithms like *k*-NN, random forest, extreme gradient boost (XGBoost), and DT method to train the data and classify it through both training and testing. Once the classification was done, the SPCDPS provides the accuracy of the model and the status of the users. The proposed model of SPCDPS architecture is represented in Figure 4.1.

4.4 Data Collection in IoT

Data gathering is the method of transferring knowledge from a variety of sensors to collect data, process the data, and find the hidden information for prediction [22]. An SPCDPS is created with the assistance of a dataset

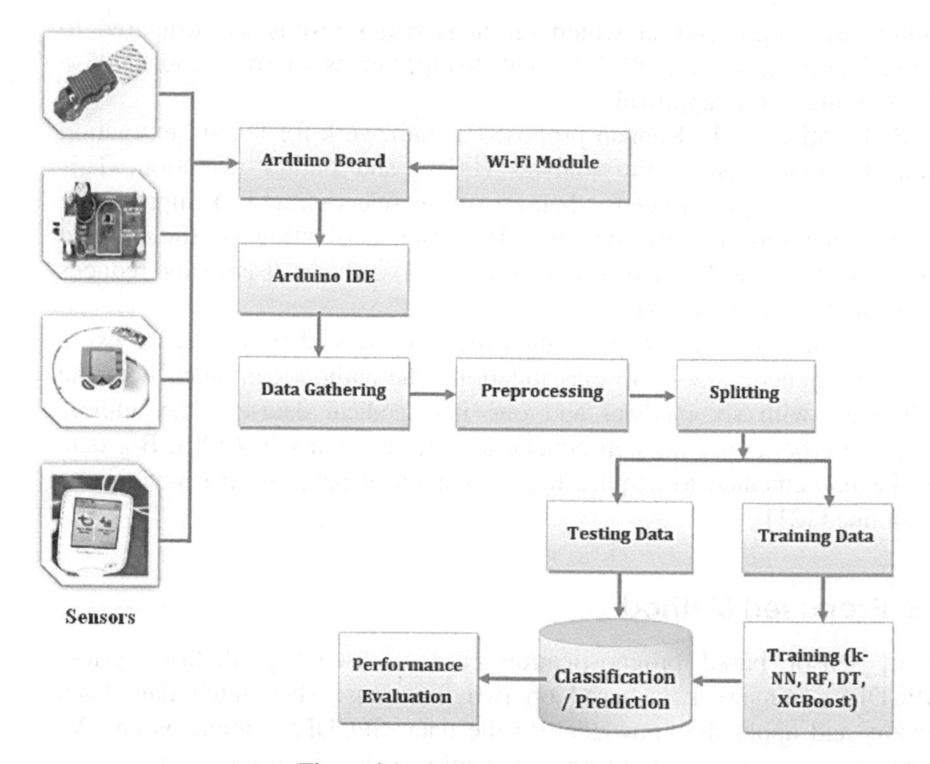

Figure 4.1 SPCDPS framework.

which can be partitioned into training and testing data. 80% of data collection is trained and the 20% data is trying to gather information that guarantees to meet its precision. To build up the SPCDPS, the dataset is collected from various users and the data is grouped into a single document and the document is converted into .CSV file for processing the data. It comprises 600 people's records and 14 highlighted characteristics which clarify the physical piece of each human. The 13 highlighted qualities and its worth are spoken to underneath with one objective variable. Figure 4.2 shows the attributes/features used in our model.

4.4.1 Fetching Data from Sensors

IoT is utilized to detect and gather the continuous information from associated gadgets, and, afterward, the data is shared over the system where it very well may be prepared and utilized for expectation or dynamic. At the

```
Data columns (total 14 columns):
 #   Column    Non-Null Count  Dtype
---  ------    --------------  -----
 0   age       318 non-null    int64
 1   sex       318 non-null    int64
 2   cp        318 non-null    int64
 3   trestbps  318 non-null    int64
 4   chol      318 non-null    int64
 5   fbs       318 non-null    int64
 6   restecg   318 non-null    int64
 7   thalach   318 non-null    int64
 8   exang     318 non-null    int64
 9   oldpeak   318 non-null    float64
 10  slope     318 non-null    int64
 11  ca        318 non-null    int64
 12  thal      318 non-null    int64
 13  target    318 non-null    int64
```

Figure 4.2 Parameters of heart disease prediction.

point when the constant spilling information is prepared as quick as gathered, it reacts to evolving conditions. Here heartbeat sensor, cholesterol biosensor, and circulatory strain sensor are used to capture the data. Heartbeat sensor is used to measure the heart rate by dispersing or ingesting it through blood. Temperature sensor is recognized in human plasma with a solitary protein. The pulse sensor estimates diastolic, systolic, and blood vessel pressure. Arduino board is used to collect the data from sensors and store it in a database.

4.4.2 K-Nearest Neighbor Classifier

K-NN is one of the most straightforward ML models dependent on supervised learning method. *K*-NN model accepts the likeness between the new information and accessible information and places the new information into the classification that is generally like the accessible classifications.

Let *X* be the quantity of working out information samples as quantitative attributes (QA). Let *Y* be an anonymous point. Store the preparation tests in a variety of information focuses arr[]. This implies every component of this exhibit that speaks to a tuple (*m*, *n*).

$$\mathrm{QA}_i = \{\mathrm{QA}_1, \mathrm{QA}_2, ..., \mathrm{QA}_n\}. \tag{4.1}$$

Calculate distance point or Euclidean distance *D*(arr[*i*], *Y*)

$$D(\mathrm{arr}[i], Y) = \sum_{i=1}^{Y} = \{\mathrm{QA}_i + \mathrm{QA}_j\}^{\frac{1}{2}}. \tag{4.2}$$

4.4.3 Random Forest Classifier

Random forest process is a controlled ordering process. Random forest is a classifier that contains various choice trees on different subsets of the given dataset and takes the normal to improve the prescient precision of that dataset. This gathering classifier fabricates various choice trees and consolidates them to get the best result. For tree learning, it fundamentally applies bootstrap collecting or stowing. For a given information, $X = \{K_1, K_2, K_3, ..., K_n\}$ with reactions $Y = \{L_1, L_2, L_3, ..., L_n\}$ which rehashes the sacking from $b = 1$ to B. Arbitrarily select "K" highlights from absolute "DS" highlights where $K << \text{DS}$.

$$\text{RnFrCs} = \text{Classifier(RS)} \tag{4.3}$$

where RS is a random sample tree. The unseen samples x^i is made by averaging the prediction $\sum_{n=1}^{N} fn(x^i)$ from every individual tree on x^i:

$$\text{RF} = \frac{1}{M}\sum_{n=1}^{N} fn(x^i). \tag{4.4}$$

4.4.4 Decision Tree Classifier

DT is a tree-like structure where data is split according to the parameters. DT consists of root node, edge node, and leaf node. Root node is selected among the attributes which has the highest entropy value. The edge/branch node is the intermediate node and it is selected based on the outcome of root node and entropy value. Leaf node is the outcome (labels) of the DT.

i. Entropy:

To construct a DT, we have to ascertain two sorts of entropy utilizing recurrence tables as follows:

1. Entropy utilizing the recurrence table of one parameters:

$$\text{Entropy}(s) = \sum_{i=1}^{N} P_i \log_2 P_i. \tag{4.5}$$

2. Entropy utilizing the recurrence table of two parameters:

$$\text{Entropy}(T, X) = \sum_{N \in X} P(N)E(N). \tag{4.6}$$

ii. Information gain:

The information gain is calculated based on before and after transformation of entropy dataset.

1. Calculate entropy of the target.
2. The dataset is then split on the various attributes and information gain is calculated as

$$\text{Gain}(T, X) = \text{Entropy}(S) - \text{Entropy}(T, X). \tag{4.7}$$

3. Pick quality with the biggest data gain as the root node.
4. A branch with entropy of 0 is a leaf node.
5. A branch with entropy in excess of 0 needs further parting.
6. The ID3 calculation is run recursively on the non-leaf branches, until all information is characterized.

4.4.5 Extreme Gradient Boost Classifier

XGBoost is an efficient implementation of gradient boosting techniques. It is one of the well-built versions of gradient boosting which is used to optimize and improve accuracy. It contains both a linear model and a tree learning algorithm. It comprises sequential learning (boosting) and parallel learning (bagging). Boosting method is a strategy that is utilized to upgrade the exhibition of the artificial intelligence (AI) model with improved proficiency and exactness. Among this, XGBoost is a propelled rendition of gradient boosting strategy that is intended to concentrate on computational speed and model productivity.

It really falls under the classification of dispersed ML community's most advanced version of gradient boosting. One of the advantages of XGBoost algorithm is its high flexibility which means we can set custom evaluation criteria and optimization objectives. Processing is faster than gradient boosting. It has built-in methods to handle missing data. Tree pruning: in gradient boosting, the algorithm stops when it encounters negative loss in the split, but in the case of XGBoost, it digs to maximum depth and starts pruning the splits without any positive gain. It is a DT-based algorithm which is considered best for small or medium structured date.

Building models using XGBoost is quite easy. But improving its efficiency is really hard. We have various parameters in XGBoost, which requires tuning. Very first, we have to create an initial model I_{m0} to predict the target PT_y as $\text{PT}_y - I_{m0}$. A new model is created based on the initial model I_{m0} with N_{m1} as a new fit. To find a fit model, we need to combine the existing model of I_{m0} and N_{m1}.

$$I_{m0} \text{ and } N_{m1} \text{ -> } F_{m1}. \tag{4.8}$$

To obtain the mean square error, MSE = $F_{m1} < N_{m1}$,

$$F_{m1}\ (A) <\text{-}\ I_{m0}(a) + N_{m1}(a). \tag{4.9}$$

Improving the performance of F_{m1} is done by *N* iteration in the fit model. The *N*th iteration is

$$F_{mN}(N) <\text{-}\ F_{mN\ -\ 1}(a) + N_{mN}(a). \tag{4.10}$$

The model should initialize with a function $I_{m0}(a)$, where $I_{m0}(a)$ is the minimizing loss function.

$$I_{m0}(a) = \arg\min{}_r \sum_{i=1}^{n} L(y_i, v) \tag{4.11}$$

$$\arg\min{}_r \sum_{i=1}^{n} L(y_i, v) = \arg\min{}_r \sum_{i=1}^{n} (y_i - v)^2. \tag{4.12}$$

Taking differential of the above equation with respect to function minimizes the mean i = 1 as

$$I_{m0}(a) = \frac{\sum_{i=1}^{N} Y_i}{n} \tag{4.13}$$

where *a* and *y* are input parameters.

4.5 Results and Discussions

The tests were intended to look at the general execution and exactness of the SPCDPS system to acquire the expectation of cardiovascular utilizing various ML algorithms. The analysis is performed against the pre-processing and feature selection scheme. Anaconda tool is utilized as a testing stage for examination. The user interface of SPCDPS framework with various attributes is shown below.

Figure 4.3 represents the anaconda software providing a platform for predicting the disease using a variety of ML algorithms. The interface helps the user to predict any kind of disease using algorithms and also it can act as

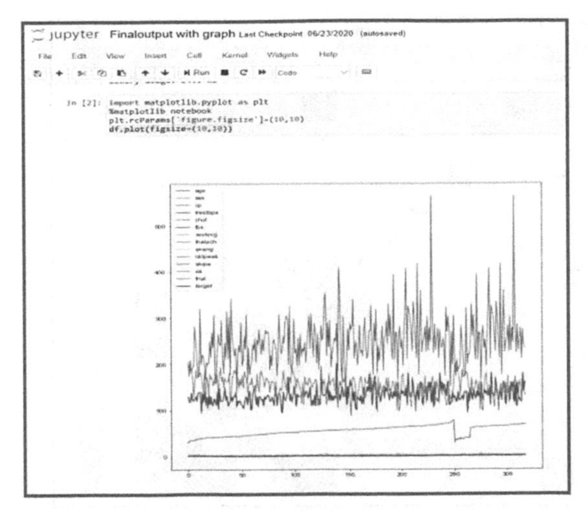

Figure 4.3 Interface of Jupiter notebook for prediction.

	age	sex	cp	trestbps	chol	fbs	restecg	thalach	exang	oldpeak	slope	ca
mean	54.981132	0.641509	1.084906	132.172956	250.081761	0.138365	0.500000	151.465409	0.267296	0.986792	1.418239	0.688679
std	10.082919	0.480313	1.036719	17.385905	56.011769	0.345827	0.507048	22.356526	0.443245	1.134579	0.628890	1.032866
min	29.000000	0.000000	0.000000	94.000000	126.000000	0.000000	0.000000	88.000000	0.000000	0.000000	0.000000	0.000000
25%	47.000000	0.000000	0.000000	120.000000	212.000000	0.000000	0.000000	140.000000	0.000000	0.000000	1.000000	0.000000
50%	57.000000	1.000000	1.000000	130.000000	244.000000	0.000000	0.000000	154.000000	0.000000	0.600000	1.000000	0.000000
75%	63.000000	1.000000	2.000000	140.000000	277.000000	0.000000	1.000000	169.000000	1.000000	1.600000	2.000000	1.000000
max	77.000000	1.000000	3.000000	200.000000	564.000000	1.000000	2.000000	202.000000	1.000000	6.200000	2.000000	4.00000

Figure 4.4 Mean standard deviation of various data attributes.

a dialogue box for animating and providing effective interaction between the user and the machine. In our dataset, there are a sum of 13 highlights and 1 objective variable.

Once the dataset is uploaded, we have to find the mean and standard deviation of several parameters. The details are listed in Figure 4.4. The technique exposed that the assortment of each highlights is different. The minimum age is 29 and also serum cholesterol (chol) is 126; likewise, based on every age group, the chol, resting blood pressure (RBS), fasting blood sugar (FBS), and other attributes vary. Thus, parameter scaling must be done through the uploaded dataset.

The correlation matrix of each attribute features is analyzed. Figure 4.5 is defined as the size of 12 $\times$ 8 by rcParams. Pyplot is used to display the

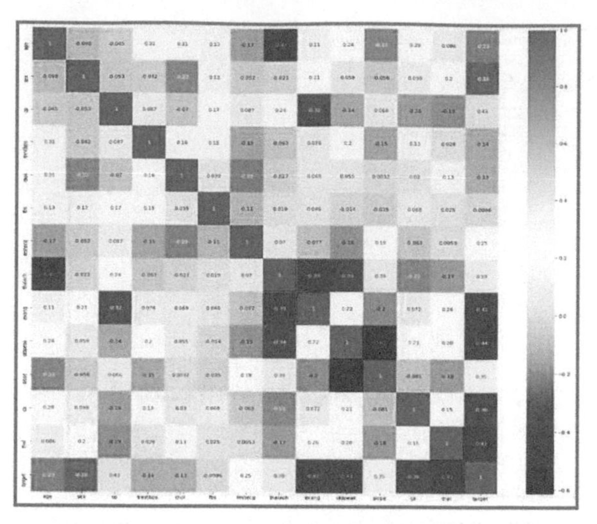

Figure 4.5 Correlation matrix for the cardiac attributes.

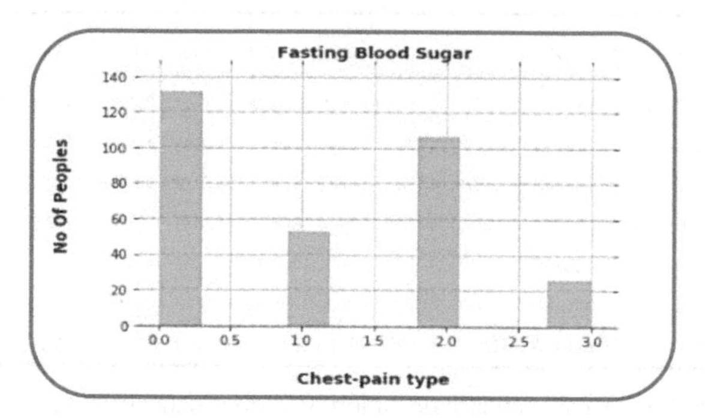

Figure 4.6 Fasting blood sugar.

association matrix. By using *x*-axis and *y*-axis, names of the attributes are added to the correlation matrix. Based on the various attribute values, the color bar is added to the matrix. There is no single component that has an extremely high relationship with our objective's worth, and a portion of the highlights has a negative connection with the objective worth and not many have a positive connection.

Figure 4.6 illustrates the attribute FBS of every user's body based on the chest pain types like typical angina, non-anginal pain, atypical angina, and asymptomatic which are shown in different ranges.

Figure 4.7 depicts that the property RBS of each client's body dependent on after effect of electrocardiogram while very still are spoken to in 3

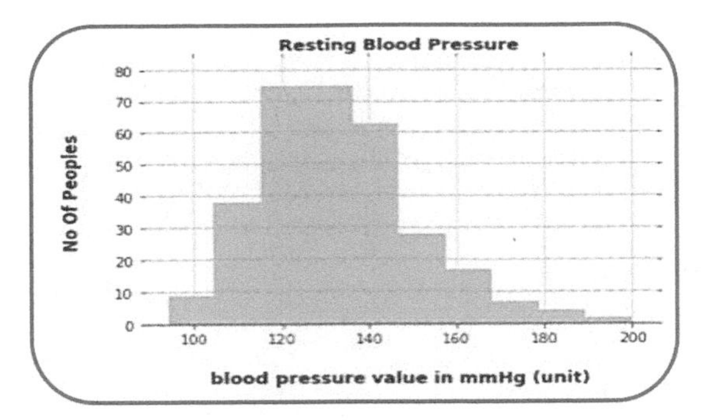

Figure 4.7 Resting blood pressure.

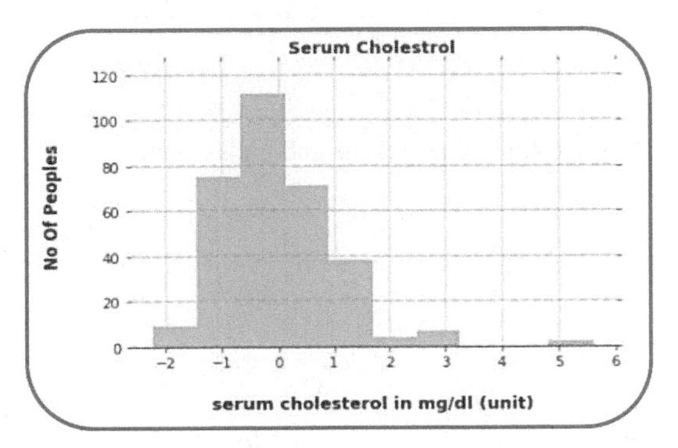

Figure 4.8 Serum cholesterol.

particular qualities: typical state is spoken to as worth 0, strangely in ST-T wave as Value 1 and likelihood or assurance of LV hypertrophy by Este's rules as Value 2.

Figure 4.8 depicts that the attribute chol is mentioned in milligram or unit based on number of users as numeric value.

To work with genuine factors, we should part each unmistakable segment into copy segments with 1s and 0s. In this, we have a component gender, with value 1 for male and 0 for female. It should be changed into two highlights, with the worth 1 where the element would be valid and 0 where it will be bogus. To complete this, we can utilize the get_dummies() procedure from pandas and we have to scale the dataset for which we will utilize the StandardScaler. The fit_transform() procedure of the scaler scales the information and we update the sections. In this, we have taken four diverse AI calculations and differed their blend of highlight and contrasted every single model. The dataset is split into 77% preparing information and 23% testing information.

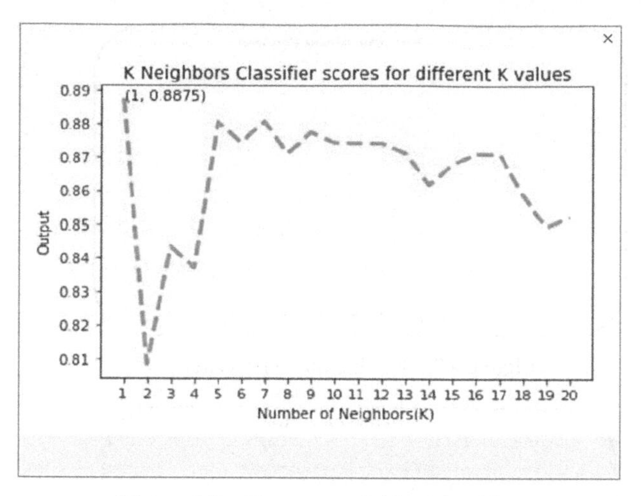

Figure 4.9 *K*-nearest neighbor classifier.

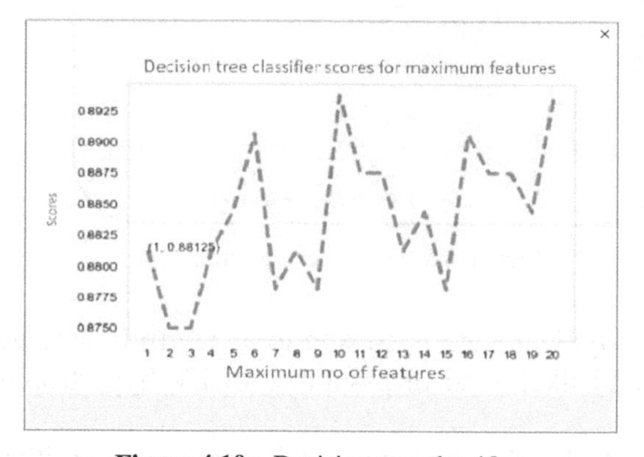

Figure 4.10 Decision tree classifier.

Figure 4.9 shows that the classifier searches for the classes of *K*-NN of a given data point and it allocates a class to this data point. Different values to *k* are assigned (from 1 to 20) and the grade is determined for each situation. Based on the performance in Figure 4.9, the *K*-NN with various classifiers gives 87.42% accuracy for the uploaded dataset.

In Figure 4.10, the DT classifier makes a DT dependent on the entropy values and classifies every data point. DT provides better accuracy as we use more features and we obtained 89.38% of accuracy for our dataset.

Figure 4.11 depicts that the XGBoost classifier creates a best fit model with low mean square value based on various sub-models. Here, we can

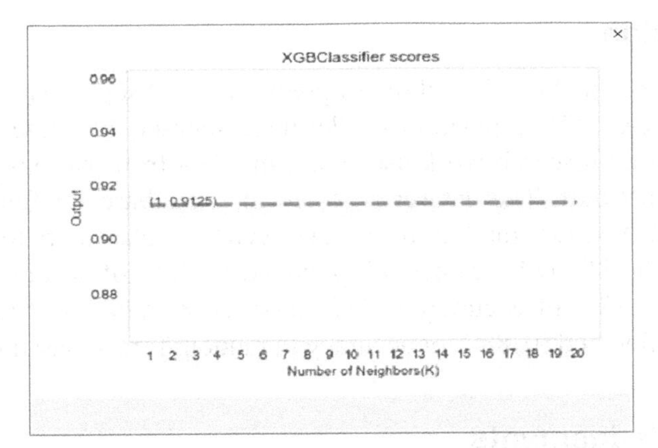

Figure 4.11 Extreme gradient boost classifier.

change the most extreme number of estimators to best fit the model and the precision of the XGBoost classifier gives a standard proportion of 89.52%.

Figure 4.12 depicts that the random forest classifier creates a DT randomly based on its various parameters to get a best tree. It assigns the parameter values of one DT to another. Combine all the smaller randomly generated sub-trees are grouped with the final one and the accuracy of the random forest classifier gives 91.88%.

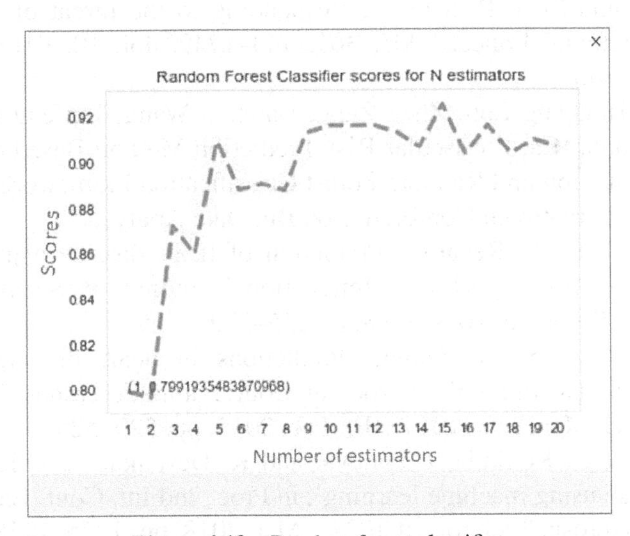

Figure 4.12 Random forest classifier.

4.6 Conclusion

Many researchers use ML algorithms to predict heart disease. The authors use data from UCI ML repositories, Cleveland dataset, or some clinical data. The proposed research work uses real-time data from various sensors using a wearable court. The dataset is stored in a database for future use. We have used K-NN, DT, random forest, and XGBoost classifiers to predict heart disease. The SPCDPS framework provided 87.42% of accuracy in *K*-NN classifier, 89.38% of accuracy in DT classifier, 89.52% of accuracy in XGBoost classifier, and 91.88% of accuracy in random forest classifier.

4.7 Acknowledgements

The authors thank the management of KPR Institute of Engineering and Technology for providing infrastructure and motivation.

References

[1] Naveenkumar M, Srithar S, Vijayaganth V, Ramesh kalyan G, 'Identifying the Credentials of Agricultural Seeds in Modern Era', International Journal of Advanced Science and Technology, Vol.29, No. 7s, (2020), pp.4458 - 4468.

[2] Ramadoss and Shah B et al. 'A Responding to the threat of chronic diseases in India' Lancet 2005; 366:1744–1749. doi: 10.1016/S0140-6736(05)67343-6.

[3] Shan Xu, Tiangang Zhu, Zhen Zang, Daoxian Wang, Junfeng Hu and Xiaohui Duan, 'Cardiovascular Risk Prediction Method Based on CFS Subset Evaluation and Random Forest Classification Framework', 2017 IEEE 2nd International Conference on Big Data Analysis.

[4] M. Durairaj and V. Revathi, 'Prediction of heart disease using back propagation MLP algorithm', International Journal of Science and Technology Research, vol.4, no.8, pp.235–239, 2015.

[5] M. Gandhi and S. N. Singh, 'Predictions in heart disease using techniques of data mining', in Proc. Int. Conf. Futuristic Trends Comput. Anal. Knowl. Manage. (ABLAZE), Feb. 2015, pp. 520–525.

[6] A. Gavhane, G. Kokkula, I. Pandya, and K. Devadkar, 'Prediction of heart disease using machine learning', in Proc. 2nd Int. Conf. Electron., Commun. Aerosp. Technol. (ICECA), Mar. 2018, pp. 1275–1278.

[7] Sanjay Kumar Sen , 'Predicting and Diagnosing of Heart Disease Using Machine Learning Algorithms', Computer Science & Engg. Orissa Engineering College, Bhubaneswar, Odisha – India.
[8] Takuji Suzuki, Hirokazu Tanaka, Shigenobu Minami, Hiroshi Yamada and Takashi Miyata, 'wearable wireless vital monitoring technology for smart healthcare', ISMICT, Tokyo, Japan, 2013.
[9] PurnenduShekhar Pandy, 'Machine Learning and IoT for prediction and Detection of stress', Haryana, India IEEE 2017.
[10] Purushottam, Kanak Saxena, Richa Sharma, 'Efficient Heart Disease Prediction System using Decision Tree', 2015.
[11] M. Chen, Y. Hao, K. Hwang, L. Wang, and L. Wang, 'Disease prediction by machine learning over Big Data healthcare communities' , in IEEE Access, vol 5, pp 8869-8879, 2017.
[12] M Naveenkumar, K Vishnukumar, 'Speech emotion recognition system using mel frequency cepstral coefficients', International Conference on Physics and Photonics Processes in Nano Sciences , Journal of Physics, ICRETM 2019, CIET, Coimbatore.
[13] Nam T. Nguyen, Nhan V. Nguyen, My Huynh T. Tran and Binh T. Nguyen, 'A Potential Approach for emotion prediction using heart rate signals', International Conference on Knowledge and Systems Engineering, Vietnam, IEEE 2017.
[14] Chauhan, A., A. Jain, P. Sharma, and V. Deep, 'Heart disease prediction using evolu-tionary rule learning', 4th International Conference on Computational Intelligence &Communication Technology (CICT), Ghaziabad, 2018, pp 1–4.
[15] Chen, H., S.Y. Huang, P.S. Hong, C.H. Cheng, and E.J. Lin, 'HDPS-Heart disease prediction system' 2011 Computing in Cardiology, Hangzhou, 557–560.4.
[16] Dewan A, and M. Sharma, 'Prediction of heart disease using a hybrid technique in data mining classification', 2015 2nd International Conference on Computing for Sustainable Global Development (INDIACom), New Delhi, 704–706.5.
[17] Gavhane, Aditi, Gouthami Kokkula, Isha Pandya, and Kailas Devadkar, 'Prediction of heart disease using machine learning', 1275–1278. https://doi.org/10.1109/iceca.2018.8474922.6. Ghumbre, S.U. and A.A. Ghatol. 2012.
[18] Shashikant U. Ghumbre and Ashok A. Ghatol, 'Heart disease diagnosis using machine learning algorithm', in Proceedings of the International Conference on Information Systems Design and Intelligent Applications

2012 (INDIA 2012) held in Visakhapatnam, India, January 2012, vol132, ed.

[19] Kavitha, R, and E. Kannan, 'An efficient framework for heart disease classification using feature extraction and feature selection technique in data mining', in 2016 International Conference on Emerging Trends in Engineering, Technology and Science (ICETETS), Pudukkottai, pp 1–5.

[20] M. Akhil Jabbar, B.L. Deekshatulu, Priti Chandra, 'Classification of Heart Disease Using K- Nearest Neighbor and Genetic Algorithm', International Conference on Computational Intelligence: Modeling Techniques and Applications (CIMTA) 2013, 2212-0173, 2013. doi: 10.1016/j.protcy.2013.12.340 - Published in Elsevier.

[21] V. Vijayaganth, P. Purusothaman, M. Krishnamoorthi, 'A Comprehensive Survey on Security Challenges and Techniques in Big Data', International Journal of Psychosocial Rehabilitation, Vol. 24, Issue 06, 2020.

[22] S. Radhimeenakshi, 'Classification and prediction of Heart Disease Risk using Data Mining Techniques of Support Vector Machine and Artificial Neural Network', IEEE 2016.

5

Assimilate Machine Learning Algorithms in Big Data Analytics: Review

Sona D. Solanki[1], Asha D. Solanki[2] and Samarjeet Borah[3*]

[1]*Department of Electronics and Communication Engineering, Babaria Institute of Technology, Vadodara, India*
[2]*Mahila Jagruti News, Vadodara, India*
[3]*Sikkim Manipal Institute of Technology, Sikkim Manipal University, Sikkim, India*
E-mail: solankisona28@gmail.com; solankiasha2710@gmail.com; samarjeetborah@gmail.com
**Corresponding Author*

Abstract

The estimated regular quantity of content created is predicted to be more than 2.5 million bytes. Furthermore, by the year 2020, it is projected that every individual in the globe will generate 1.79 MB of data each moment. Massive datasets certainly provide an enormous range of integral details, which can be used to make critical choices. Data analytics, though, needs a substantial quantity of memory and processing systems to perform efficiently. It gives the researchers and companies a fantastic possibility. This has transformed the mechanisms of analyzing and processing the enormous quantity of information, but coping with this volume of information involves optimization, and this introduces us to strategies of machine learning (ML) and computer vision. ML techniques are commonly using efficient strategies to evaluate and derive hidden details from repositories. Recognizing this tremendous opportunity depends on the ability to derive value across data collection from these huge datasets. These algorithms are predominantly used in information identification and reconstruction. This report is an analysis of algorithms such as decision tree, support vector machine (SVM), K-nearest neighbor (KNN), naive Bayes (NB), and random forest (RF) for ML. We also implement an ML technique to render massive data accessible. Besides, in the past, these models have displayed their computational speed, accuracy, and reliability. Currently, though, major problems have emerged with the dynamic features of significant information, and we encounter additional hurdles when formulating and implementing an advanced huge data management system learning methodology. Furthermore, an analysis of the evolutionary algorithms is possible to incorporate them into this different situation. These methods support to reduce the statistical treatment processing cost. This chapter illustrates the major complexities of integrating ML algorithms to the Big Data framework and demonstrates an innovative strategy to enable massive data analysis of such parameters' productivity and convenience. This chapter summarizes a literature overview of the various ML techniques. Moreover, a review is performed and discussed in this document on popularly utilized ML strategies for significant dataset insights. Implementing ML models a new methodology is utilized for scientific and health disciplines, data collection, categorization and reconstruction on large datasets that offers a solid platform and inspiration for further work in the area of data management ML.

Keywords: Machine learning algorithms, Big Data analytics.

5.1 Introduction

In technology nowadays, the quantity of content is growing at a startling pace due to advances in Internet technology, social network, mobile gadgets, and detecting. For instance, twitter generates more than 70 million of tweets every day, producing more than 8 TB every day. ABI Research estimates that more than 30 billion smart devices will be in operation by 2020 [1]. These Big Data have enormous market impact opportunities in several disciplines such as clinical services, technology, travel, Internet advertising, power management, and asset assistance. Although, when confronted with such vast numbers, conventional strategies fail. Gartner describes the concept of this massive data as vast quantity, massive speed, and more variety information that enable additional modeling frameworks to allow expertise to be uncovered, better decision making, and system enhancement.

Massive data is not defined by unique scale measurements as per this description but, instead, by the assumption that conventional techniques fail to manage it because of their scale, magnitude, or range. The concept illustrates the benefits of this huge data. Furthermore, recognition of this opportunity relies on enhancing conventional methodologies or making new, which are equipped to handle such statistics. This massive data has been alluded to as a transition for its ability that will change what they operate, function, and think. The key aim of this movement is to use vast quantities of data to allow for the exploration of information and improved choice-making. Big Data's potential to derive quality relies on information metrics; Jagadish *et al.* find analysis tools the foundation of the Big Data transformation [11].

Information analytics include numerous methods, techniques, and resources including textual analysis, market insight, visualization of data, and statistical methods, and others. This huge data is defined by five attributes: volume (collection/measurement of data), velocity (pace of data collection), variety (kinds, type, and organization of data), veracity (maintainability/quality of data captured), and value (analysis and impact). We assembled the five dimensions into a pile, which consisted of enormous scale and quality levels starting from the foundation.

The bottom part (e.g., volume and speed) is all the most hugely reliant on technological developments, while the high level (e.g., value) is far more geared to implementations that prep the main resources of huge expertise. To recognize the calculation of advanced analytics and to efficiently manage information so massive, established machine learning (ML) specifications and formulas should be modified. When challenges grow out to be gradually checking and constantly demanding resource boxes, enabling ML software enhancement fails to satisfy the algorithmic implementation needs [1].

Therefore, the conceptual accomplishments beyond boundaries will not be managed by ripping-edge registration capabilities that will enable investigators to monitor and examine a huge set of data. ML, thus, has unfathomable implications and is a key component of analysis in such huge data. Figure 5.1 demonstrates system learning application creation and depicts technologies like data science, engineering, and analytics frequently used and merged during device learning these days and its association with various streams. It may be of use to some area that requires perceiving and operating upon the information.

ML technologies have generated tremendous social ramifications in a wide range of applications, such as machine perception, voice recognition, comprehension of human speech, psychology, medical care, and the Internet of Things. It refers to the issue of how a mechanism structure can be constructed that ultimately improves across the experience. This concept

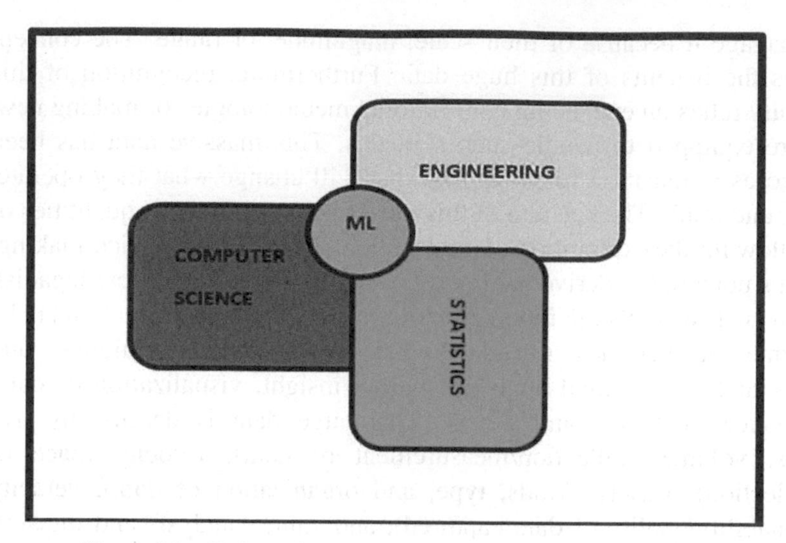

Figure 5.1 Variety of streams associated with machine learning.

is termed observing from previous practice concerning other activities and success assessment. ML algorithms allow clients to disclose a secret framework for making predictions from substantial sets of data. It descends with professional supervised learning, both abundant and large details, and efficient computational circumstances.

These days, ML is commonly utilized in the plethora of requests. Film suggestions are widely seen in Netflix, and users sorting in Gmail apps that use ML techniques. You may classify these models as described supervised learning, unsupervised learning, and learning reinforcement. With the aid of many computational criteria, this monitoring learning assigns input variables to its corresponding output. Types of techniques that use supervised learning are classification and regression.

Classification is meant for forecasting distinct valued variables, while regression is employed for constant valued characteristics. Applications of classification techniques are the decision trees (DTs), support vector machines (SVMs), naive Bayes (NB) Algorithm, etc. Regression systems are usually employed for linear and lasso regression [2]. Frameworks that employ supervised learning techniques are fingerprint attendance, weather forecast, report classifier, etc. Unsupervised learning from the specified database recognizes trends.

Grouping, interaction, and the elimination of fractal dimension are instances of unmonitored learning. Clustering methods are popularly seen in *K*-means, K-Fediod, fuzzy mean, CLARA, DBSCAN, and OPTICS. Artificial neural network (ANN) is a strategy related of being stuck in local optimization, both supervised and unsupervised, culminating in slow convergence. Proposal for Amazon item is a typical approach using unmonitored learning methodology.

Learning to strengthen research by discovering the quest setting and spatulas on strike and check process Reinforced learning is discovered by analyzing the search framework and functions on trial and error process.

There are certain specialized algorithms in ML, i.e., deep learning, virtual learning, incremental learning, representation learning, etc., that are growing significant prominence these years and are quite beneficial for Big Data analytics. Deep learning is the ANN influences at MLT that means Machine Learning Techniques [3]. It offers a centralized portrayal of the information in which high-stage characteristics can be characterized in terms of functionality at low rates. It can be utilized in, respectively, monitored and unmonitored learning and is capable of learning from unidentifiable sets of data.

The McKinney International Organization has said ML would be one of the significant information uprising's key drivers. The justification for this is their chance to recognize from information and give ideas, choices, and forecasts motivated by detail. This is focused on estimates and can derive patterns from data similar to statistical analysis; however, it does not require the extensive use of empirical data. The two primary categories of learning functions, as per the design of the existing data, are supervised learning when all inputs and their expected output (labels) are known and the model knows to link inputs to outcomes and unmonitored learning when desired outcomes are not identified and the mechanism is not recognized detects the framework within the details itself.

This chapter concentrates on ML as a cornerstone of information analysis. Specifically, we mean researching its opportunities and challenges on Big Data that exposes frameworks for ML. Big data, for instance, entrenches multigranular architecture learning and assorted variety, similarly from multiple perspectives. It gives possibilities for correlation depreciation in the context of sequential chains. Massive data also provides ML significant challenges in several cases in extracting ideal structure from accessible testing data.

5.2 Literature Survey

Numerous works are available in literature in this domain of research. Selective technically influential works have been picked-up for a discussion in this section. J. Qiu *et al.* implemented various ML techniques for the analysis of large data [12]. The first is description or function learning that interacts with depictions of ensemble learning that facilitate the phase of information processing. The efficiency of the machine evolutionary algorithms is considered to be heavily affected by the choice of statistical analysis (or features). This arrangement of training plays a vital role in the activities of minimizing complexity. The major components within classification tasks are primary attributes, an abstraction of characteristics and proportional range learning. Technologies for identifying characteristics (differential choice) are being used to determine specific characteristics of data that are very significant for design development use. Methodologies for the abstraction of functionality turn huge-dimensional information into a domain of smaller size. A gap feature is built-in remote data augmentation to quantify the range between the different stages of a set of data.

The investigators listed in their paper regarding another strategy of hot training, called deep learning. Many of the traditional strategies to system learning implement a deep-structured learning framework that includes a single sheet of various representations in software. Instances of such learning techniques comprise Gaussian mixture models (GMMs), hidden Markov models (HMMs), SVMs, logistic regression, Kernel regression, etc. Unlike the shallow-structured learning design, in deep architecture, deep learning techniques employ monitored and unmonitored techniques.

The training structures with a deep processing framework are constructed of many spatial simulation phase layers, wherein the outcome within each lower stage is supplied as the source of the higher subsequent stage. Instances comprise deep learning networks, traditional neural networks, deep networks of convictions, recurrent neural networks, etc., and these are well equipped for Big Data analytics implementations because of the outstanding quality of deep learning algorithms.

Interoperability with the conventional ML algorithms is a demanding task. The conventional strategies cannot manage the enormous collections of information within a prescribed period because they involve all the details in the same server. To overcome this issue, a new area of system learning has emerged, which is termed as distributed learning. In this process, research is performed on sets of data dispersed between many workstations to improve the research procedure. Interpretations of the techniques for decentralized ML are laws of judgment, layered generalization, meta-learning, dispersed boosting, etc. Simultaneous system learning is yet another important instructional technique wherein learning is performed between parallel computing conditions or on numerous synchronized machines.

Transfer learning is another methodology for ML noted in their paper. In the traditional ML mechanism, standard practice is to take both the training information and analytical results from a similar sector. It means the domain of the input method and the transfer of information is similar. Yet, there are some other situations where it is a complicated and costly job to obtain learning and trial results from a similar field. The transition learning methodology has been used to fix this problem. In this system, training people from a similar target domain for a source domain generates a highly efficient learner. Transfer learning models are used extensively in many systems for the analysis of information in the actual globe.

The writers mentioned another method of learning, called effective learning. For certain instances, lacking marks, the details are portrayed which becomes a problem. Equating this significant amount of data manually is a costly and challenging process. It is very tough to learn from unidentifiable details, too. Active learning is employed to tackle the prevailing problems by choosing a subset of the most adequate identifying instances. Another technique, kernel-based learning, has been extensively utilized for the construction of effective, functional, and highly scalable variation algorithms in several engineering applications. Plenty of the kernel-operable architectures are SVMs, principal component analysis (PCA), perceptron cores, etc.

A work detailing the commonly deployed ML algorithms for Big Data management was proposed by J. L. Berral-Garcia [13]. Numerous methods are included for forecasting, estimating, and convergence functions. This document introduces the DT algorithms (such as CART, recursive partition trees, or M5), K-nearest neighbor (KNN) algorithms, Bayesian algorithms (using Bayes' theorem), SVMs, ANN, *K*-means, DBSCAN algorithms, etc. Numerous implementation structures are also listed: MapReduce Frameworks (Apache Hadoop and Spark), Google's Tensor Flow, and Microsoft's Azure-ML. Architectures of the subsequently mentioned techniques are made publicly available across various platforms, methodologies, and modules such as R-cran, Python Sci-Kit, Weka, MOA, Elastic Search, Kibana, etc.

M. U. Bokhari *et al.* proposed a three-layered design of structure for huge data processing and analysis [14]. The three layers are layer for collecting data, analyzing information, and for creating statistical analysis and recording. A collection of high-velocity networks or database is placed in the collecting data system to obtain and manage the massive amount of big information classified as high-intensity outlets such as detectors or social networking sites. This is the information-processing unit that holds this massive data. Data management can be achieved using the Hadoop Distributed File System (HDFS). ML techniques such as ANN, NB, SVM, PCA, etc., are utilized in the data processing system to generate information from the enormous dynamic sets of data.

In their paper, P. Y. Wu *et al.* presented methodologies to demonstrate how advanced analytics was beneficial in precise healthcare, presenting those individuals with the most effective treatment [15]. PCA, singular value decomposition, and tensor-based techniques are helpful for the abstraction of functions, and wrapper-based approaches are effective for the interface collection process. Both of these are strategies for reducing complexity. The paper compares the different methods used to conduct data mining algorithms. Logistic regression, cox regression, and the strategies of local regression are easy to understand but are vulnerable to anomalies. Logistic regression with the formalization of LASSO (Least Absolute Shrinkage and Selection Operator) decreases task storage. However, there is a concern with alignment. Often necessary for making information mining techniques are other systems such as HMMs, conditional random fields, hierarchical subtype exploration, event principle extraction, etc. The researchers spoke about the valuable advanced analytics applications. Apache Hadoop, IBM Info Sphere Cloud, Apache Spark Streaming, Tableau, QlikView, TIBCO Spotfire, and certain data visualization platforms are extremely effective systems for developing approaches for huge data exploration. Two real-world research papers like interdisciplinary bioinformatics information to determine grasp leukemia processes and the implementation of genomic information into the EHR program to enhance clinical care and prognosis were performed to explore the utility of medical analytics for targeted therapy. In this analysis, multi-omic TCGA (The Cancer Genome Atlas) data and EHR (Electronic Health Records) data were utilized.

M. R. Bendre *et al.* performed a report on the application of data analytics in sustainable farming [16]. The investigators referenced that large information offers a wide range of activities for uncovering fresh perspectives to resolve different agricultural issues. The developed structure uses the strategy of MapReduce to analyze data and the methodology of linear regression to forecast the results. Information gathered from the platform KVR (Krishi Vidyapeeth Rahuri, Ahmednagar, India) was considered for validating the system. The outcome predicted applying this method is quite beneficial for productive policy taking in the farming sector.

5.3 Big Data

Numerous scholars worldwide extensively use the terminology Big Data that represents an incredibly vast information set. It is a standard concept used to represent incremental growth and the accumulation of unstructured and structured details. Several sectors from criminal justice to immovable properties and medical services have embraced these analytics and achieve several fold profits. Industry's core requirement is data that can be actioned. Company outcomes from the review and indicates strategies. The functionality involves monitoring it from violence to climate, to purchasing, and products [4].

The differentiating element here is the potential of giant information to deal with large volumes of unorganized actual-time data. Together with system learning, it assists to make a significant effect on the performance of operation and client service. This is a fantastic distribution to many other end users of details and expertise from the devices. This is an outstanding origin of our processes and customer's awareness and data. The Big Data movement intends to modify how we operate, function, and communicate by allowing system automation, encouraging exploration of knowledge, and enhancing policymaking. This is a thriving software technology development field with several other industry verticals [6].

It has achieved tremendous popularity in large and diverse sectors of operation. Those comprise social networks, economics, banking, education, farming, etc. Many smart ML technologies have been implemented to offer approaches for the statistical insights of huge data. This happens not only with growing data volumes but also with dynamic and diverse computing and assessment process. Gigabytes of information are now being formed regularly for days that might acquire features such as large velocity, big quantity, instability, non-linear information, real-time data, etc.

It has drawn considerable support from medical care, data science, and information systems scholars. As a response, data output would be 44% higher at this point than in 2009. Thus, the size, speed, and images are classified and rise substantially. Utilizing conventional technology, it is expensive to retrieve, analyze, and envision these large datasets. Large data and ML are in certain form transforming almost all facets of everyday life.

Netflix recognizes what users like to see films and Google learns what they desire to find depending on their browsing history. Google has progressively started to introduce most of its current non-ML systems with ML methods, and there is considerable hope that significant trends will be made in other fields. Besides, other companies such as Twitter, LinkedIn, and Facebook utilize predictive analytics in the social media environment for specific use instances. Furthermore, applications of these use scenario frameworks have been released around the globe.

However, a semantic framework has been restricted for the particular implementation of technology. Classical information using the unified structure of the server is implemented where a central software device addresses complex and diverse issues. Centrally controlled software manages vast volumes of data expensively and inaccurately. Big data is focused on the structure of the hierarchical server, where a massive data frame is handled by breaking it into many tiny pieces. Thereafter, numerous separate machines existing in a defined communications network simulate the remedy to a query. It is one of the emerging innovations as it offers the best results from large and diverse information.

This contains deciding the correct broad information storage and operational structure, supplemented by efficient ML models. Due to the substantial excitement encircling significant information processing and its benefits, there has, thus, often been no comprehensive research study fixated on simultaneous information excessive system learning algorithms for data collection. The machines interact among themselves to develop a way to an issue. Contrary to the consolidated database server, the distributed system offers efficient storage, reduced cost, and enhanced productivity [4].

It is since a centralized system that is focused on mainframes that in a distributed system are not as affordable as microprocessors. However, the distributed database has more processing resources relative to the centralized database structure, which is utilized for conventional data processing. Standard relational servers cannot manage Big Data. Immense quantities of sets of data are hailing from various outlets such as detectors, relational software, Internet, online networks, etc. This concept can be represented by analyzing the various “V”s correlated with them: Volume, Velocity, Variety, Veracity, and Value [7].

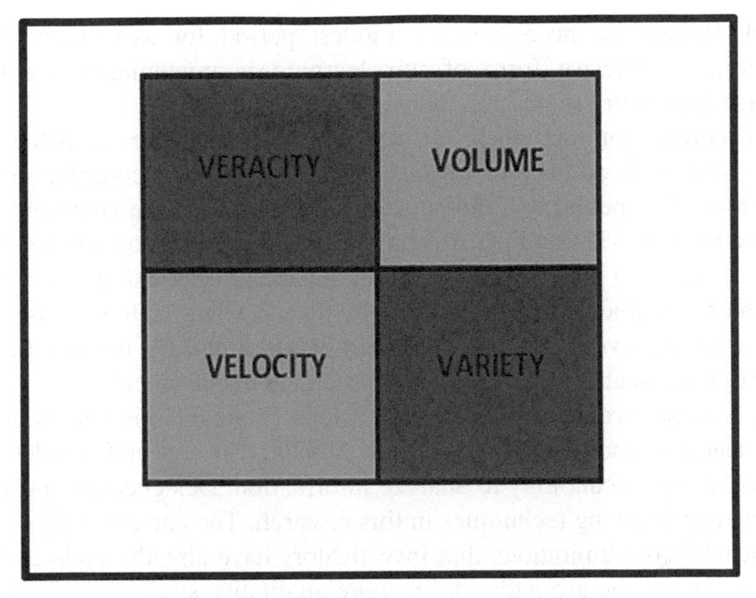

Figure 5.2 Correlation between various "V"s.

- Volume: This signifies the massive quantity of information that is generated each minute, fluctuating to megabytes between gigabytes. Utilizing centralized structures, these large quantities of data can be retained.
- Velocity: This term reflects the speed at which energy is generated and handled to aggregate requests.
- Variety: It shows the variety of data we will use.
- Veracity: This refers to the durability of the data. That is, the biases, distortion, deformity, etc., in the data are indicated.
- Value: This indicates the valuable expertise revealed in the data. Data analysts take benefit of several excellently marked methodologies. Figure 5.2 indicates the correlation between Veracity, Volume, Velocity, and Variety.

Content analysis, predictive modeling, speech recognition executing, ML, etc., are some strategies for picking large datasets quicker and better to discover secret perspectives. There is a necessity for an innovative phenomenon articulated by huge data to examine huge amounts of data, accumulate technologies, related systems, and conduct required analysis on huge data frameworks. This technology differs throughout complex devices and arrays of single-purpose subdevices. The information recorded from multiple sources necessitates a modest period for assessment and organization. A cohesive form of this learning is implemented on the MapReduce framework to possibly boost up the development.

In all corporation activities, an extensively accessible MapReduce computing method is adopted to various learning techniques contributing to the ML group. The period and information processing will be enhanced by using these models with Hadoop for massive storage allocation. Hence, we ought to evaluate Big Data appropriately in the actual period to produce the outcome more reliable. One reason for doing that is using HDFS. However, the analysis of massive data can be achieved by utilizing ML. It implements science-based and technology-suspicion-based data management.

There is often diversity in the growth at UIN (State Islamic University), its varied learners, and their thought about Muslims. To get updated details on that, we require technology to analyze information. Designers are indeed discovering new learning techniques in this research. The current study is an effort to identify the limitations that investigators have already made in the study, hence laying the groundwork for more qualitative studies. It offers an insight into the collaboration between ML and Big Data collection.

5.4 Machine Learning

ML is an extensive research sector that incorporates insights from various engineering categories, notably, artificial intelligence, statistics, the theory of information, algebra, etc. Its research focuses primarily on building rapid and effective learning algorithms that can make data assumptions. It is at its center due to its potential to discover from information and offer perspectives, choices, and forecasts data-driven. Nevertheless, the management of this body of data requires optimization, leading to a development in data analysis and ML.

It is a methodology employed when interacting with information modeling to construct predictive models. It usually involves phases of segmentation, processing, and evaluation of knowledge. Before the processing of data, available coarse details into the "correct shape" for subsequent learning phases are produced. Certainly, the actual data will be unorganized, raucous, segmented, and in conflict. This step transforms such data into a frame that can be used as studying contributions via data cleaning, retrieval, alter, and confluence [2].

The testing phase selects formulas for research and guides display specifications for generating desired outcomes using the data input

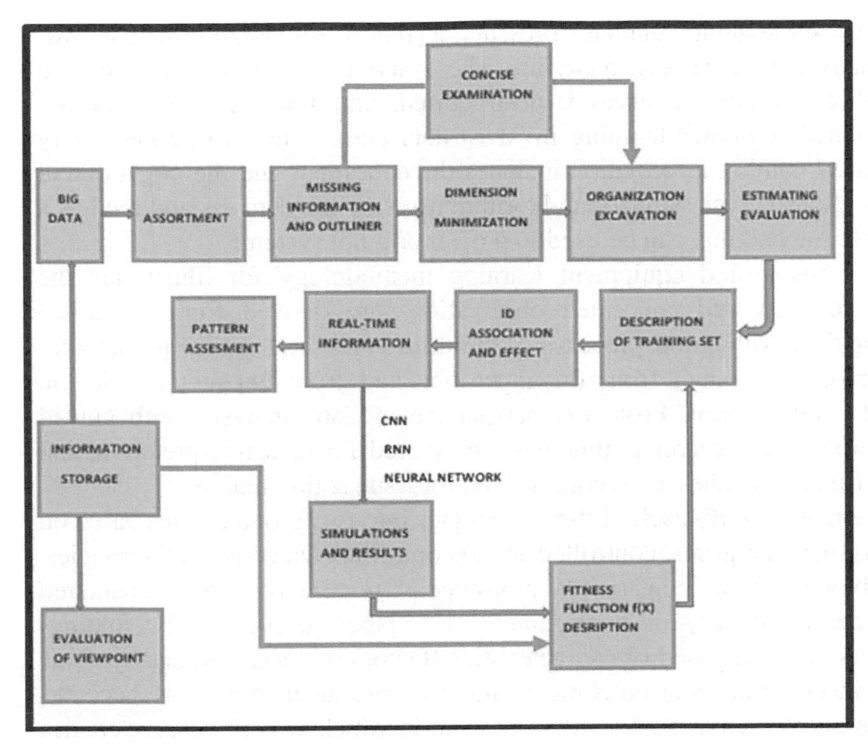

Figure 5.3 The involvement of ML in Big Data.

precompiled. A few learning methodologies, notably insightful learning, can also be used for preprocessing details. The evaluation decides to implement the trained models afterward. For instance, a training dataset's performance evaluation requires a set of data assessment, performance calculation, error estimate, and empirical testing. The outcomes of the study can cause modifications to the specifications of the selected learning measurements and select various calculations in turn.

Constructing such algorithms comprises four selection of architecture: 1) selecting details for the learning; 2) deciding a destination function; 3) preferring performance; 4) deciding method for learning. A few of the current algorithms have been introduced for sets of data that could fit entirely into the storage. While the information keeps through day by day, many smart techniques of analysis have been incorporated to offer approaches to multiple challenges with massive data statistical insights. Figure 5.3 illustrates the involvement of ML in Big Data.

The subsequent segment presents a report on some effective ML approaches for advanced analytics. These activities are generally divided into three groups, supervised, unstructured, and practicing, to improve. Supervised computer learning involves data classifying instruction. Every annotated training information includes the data input and the target output for the goal. The classification algorithm assesses the datasets and renders a variable implied that can be used to project original systems.

In unmonitored equipment learning methodology for illustrating the batch research and concealed observations are derived from unmarked sets of data. Validation processing, the third category, enables a computer to derive its conduct from the input obtained from encounters with an outside environment. From the perspective of data analysis, both guided and unprotected teaching strategies are favored for data interpretation, and validation approaches are recommended for issue action making.

We may identify each of these strategies into two broad classes based on our simulation reasons: controlled and uncontrolled. On controlled strategies, the objective is to construct a prototype *a posteriori* from encountered instances to identify or approximate class labels *a priori*. The findings (aspects) are composed of characteristics (factors or variables) and symbols (dependent variables or outcomes), and the consequent framework becomes a component such as $f(x1 \ldots n) = = 1 \ldots m$ in which xi is the characteristics and the logos generated [5].

A design is adequate if the variation between the symbols created and the actual associated metadata in the sample instance is minimal. For various modeling optimization techniques (ultra-variables), we can enumerate the simulation model until we are delighted with the proposed method, but not until ultimately evaluating the identified system with observations not used during the adaptive practicing. Recognize data gathered in network infrastructure with VMs providing Internet services, for say we ought to be able to accurately predict how often workload a VM will create by analyzing exactly how many customers parsing the system.

We ought to recognize data gathered in network infrastructure with VMs providing internet service to accurately predict how often workload a VM will create by analyzing exactly how many customers parsing the system.

As the system supplier does not have the authority to audit their user VMs, he/she is gathering samples of the number of transactions and packets against a user VM and the request needed by the VM. The supplier then utilizes a logistic regression technique to produce a framework that will anticipate the request that the VM would need if that circumstance arises, delivering a set of slots and a data packet against the VM, and offering it before breaching the negotiated SLA or oversupplying assets.

On unmonitored research, we aim to construct a framework from observable instances that inform us what the information corresponds to, which means we do not have symbols for our information and we want the system to inform us the tags will adequately classify our data. The method is identical to be monitored, however, without illustration outcomes because our model's approximate outcomes are the relevant information that we are searching for. Assume, for example, data gathered from an Internet traffic report, and the ISP would like to learn how its customers reply to a particular collection of activities or trends.

The ISP gathers traffic illustrations and then relates a scheduling technique that offers a method that discriminates customers by traffic conduct, melding them into *k* groups, each group sharing significant conduct. The ISP administrator can then check the features of each community and, if content with the data is collected, use the design to tag new customers employing the quite appropriate group formed to mark and apply various guidelines per each category due to their specifications.

5.5 File Categories

Conventional server structures are built on standardized data, i.e., conventional information is contained in a file or a predefined sequence. Defined data instances embody relational database system (RDBMS) and documents that only address queries about what occurred. The classical repository offers only a limited-level perspective on an issue. However, to increase a company's effectiveness, it is used to obtain more visibility into the details and to learn about unorganized metadata. It ideally incorporates unstructured or semi-structured information that strengthens the data collected from various sources, such as clients or viewers. It converts it into knowledge-based details after the gathering.

5.6 Storage And Expenses

It is very costly to manage large quantities of data in the conventional server system because not all the information can be retained. It will reduce the quantity of data to be evaluated that would diminish the consistency and reliability of the outcome. Then the data is processed in huge data structures and the connection points are established which will include a significant level, although the volume designed to process huge volumes of data is smaller in massive data that gives more appropriate outcomes.

5.7 The Device Learning Anatomy

Handling tremendous data usually entails identifying now on the critical details to demonstrate the rust off, change the quality, or condition by prolonged representation to the data that generates it, and modifying it to suit data and information. The dominant portion of undeniable and prophetic estimations is extraordinarily beneficial for these targets. ML approaches to illustrate and predict, consortium and agglomerate details, and disclose details.

It is as a portion of data managing provides strategies for systematically manipulating and concentrating information from research, whenever individual managers and specialists are reluctant to reach due to the level of multidimensional existence or the volume to be worked with per measurement. It has been a science linked for a while in monstrously particular cases such as medicines, atmospheric sciences, or providing; interestingly, in the approach of the perception of huge data, everyone else has the precondition for classifying it and further gaining structurally from it.

As a response to the observed, countless phases, systems, languages, and implementations have popped up to classify that data were offering ML computations a few to stable prepping, a few to retrieve fantastic instances in data, some for distributing circumstances, but all extending the still emerging hole between AI and investigating and swapping [8].

However, such tools are crucial viewpoints for linked types of ML, apart from the accuracy and multidimensional existence of computations: perspectives including the potential to balance planning and anticipation, the sort of coding language to be employed to illustrate the downfall and restore the overview, accessible formally enforced databases, stable preferable, or method results active region evaluate intended to be collected and demonstrated in this designers want to introduce protection on the concepts of machine data collection and data preparation, and numerous software and phases used among data scientists to consider massive data and coordinate ML aspects.

Apart from the accuracy and multidimensional existence of calculations, such systems are having crucial viewpoints for associated ML model. The perspectives include the following criteria: the potential for balance planning and anticipation, the sort of coding language to be employed to illustrate the downfall and restore the insight, stable and accessible enforced databases, method that evaluate active region, results intended to be collected and demonstrated in this module to introduce protection on the ML data collection and preparation, numerous software and stages utilized by data scientists for massive data and to coordinate various aspects of ML.

5.8 Machine Learning Technology Methods in Big Data Analytics

In Big Data cases, subscribers, administrators, and analysts in data ought to obtain data and know from vast sets of data and massive data spikes. Recognizing the ultimate aim of building this technique efficiently, enabling data investigators to focus on manufacturing this process and concentrating on the results, a few structures have appeared to be offering such operation. A few of them, though in no way the only ones, are abbreviated here.

5.9 Structure Mapreduce

Apache Hadoop and Spark although many device-learning methods can be provisioned by comprehending a computation process for each data and then the entire method generates in the configuration after that. In Apache Hadoop and Spark, many device-learning strategies can be implemented by analyzing a calculation process for each data and later that entire method generates in this configuration.

These processes can be clarified using a MapReduce method where information is divided into sections, every component is analyzed in turn and the results are accumulated in the structure.

Apache Hadoop is an accessible source structure widely used for such activities in Java. Customers can send piles or data centers without anybody else, or they can get a response from companies that offer Hadoop as a portal as a provider and concentrate solely on presenting their implementations. In Hadoop's view, Apache Spark is a response that seeks to strengthen implementation by focusing on features such as ML, chart assessment, and information broadcasting, and computerized in Scala or Python, in specific. Although Hadoop has often been accessible but now has numerous resources serving more sections of the industry, Spark and Mahout evolve by concentrating and optimizing unique problems, a significant number of them connected to ML and tremendous data processing.

5.10 Associated Investigations

ML activities involve supervised learning techniques, unmonitored learning techniques, and validation methods. Unmonitored methodologies simply begin from unmarked sets of data, so in a manner, they are linked to discovering unidentified properties (e.g., samples or guidelines) in them. It concentrates on prediction, focusing on recognized characteristics acquired from training samples [8]. Data mining (that also is the skill enhancement in datasets step in evaluating) concentrates on the exploration of (originally) unseen data features.

For example, a classifier's achievement assessment includes given data collection, efficiency quantification, fault evaluation, and qualitative testing. The outcomes of the study contribute to the adjustment of the specifications of the selected learning parameters and/or the selection of various iterations. While effective implements of ML will not ultimately depend on frequently unloading bulk data observations in computations and searching for the finest, the ability to use a bunch of data for ML activities is now a propositional necessity for experts.

Although most ML remains true irrespective of data volumes, some elements are the unique realm of massive data processing or that relate to limited quantities of data than others do. The method entails routes and computation for illustrative, statistical, and didactic analysis. Pertinently, this method is specifically stated as recurrent, which can be particularly essential for analyzing the massive amount of data, and disintegrates the total amount of data at each step of the design process of a learning object.

Such optimizations are defined by knowing a reference feature (f) that masks the appropriate input parameters (X) to a variable output (Y); $Y = f(X)$. This is a general cognitive activity, where we want to make additional (Y) forecasts, given different input parameter (X) instances. We may not know what the component (f) or its structure appears like. If we did, we will use it explicitly, and using such approaches, we would not have to understand from the information. Learning the modeling $Y = f(X)$ to create forecasts of Y for the latest X is the most prevalent subset of ML. It is called security analysis or prescriptive analytics and the key goal is to create the forecasts that are as reliable as possible. If we utilize huge data to store massive quantities of information and deception, one point is, but it will be feasible through system learning, to retrieve helpful data from them. This learning enables us to derive productive correlations.

5.11 Multivariate Data Coterie in Machine Learning

To assess an optimization in this learning, the quality estimate is very much essential. The unidentified sample data efficiency measure is known as a testing dataset that is dissimilar from the set of data. It is called the set of data that teaches itself as a training dataset. Data collections are developed from a probable allocation, in a fair scenario. When generating the sample, we usually agree that the understanding in the dataset is likewise endogenous from one another that they are dispersed imperceptibly. This implication is normally referred to as autonomous and indistinguishably distributed (a.i.d). The training and test set of data for the measurements essentially must emerge from the same allocation of possibility and are a.i.d. After dividing our information into training and test sets, we organize the learning computations using the training set to decrease the training inaccuracy. At that juncture, we are assigning the "trained" method to the test set, which are unremarkable details. Repeatedly more prevalent is the test error than the training error. This means that we will reduce the training error and further lessen the difference between both the training samples and the testing mistake.

5.12 Machine Learning Algorithm

5.12.1 Machine Learning Framework

This domain is directed at exploring models and strategies for automating the processing of new insight, advanced capabilities, and innovative forms of generating established data. This part addresses specific ML methods within the sector.

5.12.2 Parametric and Non-Parametric Techniques in Machine Learning

Presumptions can significantly alleviate the problem of learning but they can also restrict whatever can be understood. These are called parametric ML algorithms that compress the feature to a defined structure. The implementations require two stages: 1) for the task to pick a type; 2) using the training details to determine the constants for the task. The linear and logistic regressions are a few illustrations of parametric ML methods.

Non-parametric ML algorithms do not create specific claims about the type of scaling factor. They are allowed to acquire any current design from the training results, by not producing hypotheses. Often non-parametric techniques are more versatile, achieving greater precision while demanding much more information and training period. Applications of such architectures cover vector systems, neural systems, and DTs.

5.12.2.1 Bias

They are trying to simplify predictions that a design creates for better learning of the desired task. Parametric algorithms usually have an elevated bias that makes them simple to master and simple to implement but typically less versatile. In addition, they provide slower forecasting accuracy on complicated situations that decline to satisfy the interpreting application bias predictions. DTs are an illustration of a minimal bias application, while linear regression is an instance of a method with elevated bias.

5.12.2.2 Variance

It is the extent that the objective feature estimation would alter if various training results were employed. This method forecasts the objective act from the training samples, so we must anticipate that the classifier does have variability, not null deviation. An illustration of an increased variability technique is the KNN methodology, while linear discriminate analysis is an instance of a reduced variability method. Any computational modeling system or learning system has the objective of achieving low accuracy and lower value. The objective of any computational modeling ML system is attaining low accuracy and variance.

The method, in addition, will create a better output on estimation. These applications' lightheadedness is always a struggle to manage bias and variance. Through growing the bias, the variability will reduce. Expanding the variation diminishes the bias. Predictive analytics is mainly about reducing a system's faults or producing the greatest correct estimates feasible at the cost of illustrating capability. We are going to lend, recreate, and clone techniques from various areas, namely analytics, as well as use them for those reasons.

5.12.3 Parametric Techniques

5.12.3.1 Linear Regression

A method represents a sequence that better matches the correlation among the source parameters (x) and the desired output (y) by determining different weights for the parameter (B) response variable.

$$y = B0 + B1 \times x. \tag{5.1}$$

In consideration of the inputs x, we can estimate y, and the objective of the linear regression supervised modeling is to identify solutions for the constants $B0$ and $B1$. Numerous strategies can be used to develop, from details, this analysis, such as a linear algebra approach for regular minimum ratios and differential modeling [10].

We have enough than two groups, so the linear discriminate assessment methodology is the recommended linear classification model; a few excellent thumb norms while using this methodology is to eliminate equivalent (associated) factors and, if feasible, to disable distortion from the statistics. It is a rapid and effective methodology and a great first automated system. The LDA (Linear Discriminant Analysis) description is constructed of statistical information attributes, which is determined for every category.

It involves the average scores in each category in the overall group for a specific input parameter of quantified variability. Probability is calculated by measuring a discriminating valuation for raising category and predicting the highest values for the group. The theory predicts the information has a Gaussian distribution (normal distribution) and extracting anomalies from the results is a wise decision. It is a convenient and straightforward approach for predictive analysis classification tasks.

Supervised architectures for computer learning are those implementations that require special help. The collection of source details is divided into sets for training and testing. The training results' collection has a performance vector that needs to be determined or assessed and regressed. Most techniques discover a kind of structure from the test set and implement them to the forecasting or categorization of test input data.

5.12.3.2 Decision Tree

A categorization tree is a forecast, f: x to y, which determines an example-related label, x, by moving from a tree's source node to a leaf. Some of the finest-supervised techniques for learning are commonly recognized as a DT. This produces a graph or tree that shows the possible result of an action using a branching methodology. Inner node tests a feature in a depiction of a decision node branch that correlates to the outcome of the parent module, and, eventually, every leaf applies the class name. An up-below approach is being used to categorize an issue, beginning at the base of the tree. For a given component or module, the group that correlates to the sample point's valuation for that characteristic is regarded until a leaf is attained or a tag is determined. In Figure 5.4, the articulation of ML algorithms is described.

5.12.3.3 Naive Bayes

It is a popular methodology, exploiting the likelihood of the situation. There is a possibility table in this method and this column is focused on feature vectors that ought to view the possibilities of the category to predict a new data point. The basic idea is that it is contingent autonomy, and it is assumed "gullible." Most input attributes that are separate from each other will probably hold valid. This optimization renders massive data structures extremely simple. It is a powerful identification tool though. The Bayes derivation offers a baseline for subsequent equations of probability [14]. It is as indicated in Equation (5.2).

$$P(clx) = \frac{P(xlc)P(c)}{P(x)} \tag{5.2}$$

5.12.3.4 Support Vector Machine

Vapnic presented a methodology named SVM in 1995. It utilizes a distinct hyper-plane or describes the borders of actions across a set of instances categorized with different titles. An ML model is famous; it can be an organization of its own. SVM is a properly supervised technique for identification. The technique requires learning technology to analyze an ideal hyper-plane and this explains the choice and, in turn, categorizes the latest illustrations. It can conduct both static and non-linear categories, depending on the kernel in implementation [5].

5.12.3.5 Random Forest

Random forest (RF) architectures are a proprietary concept for an assortment of DTs. RF is indeed a collection of DTs. Categorize a new organization into various attributes and revert for that class by recognition and votes. The forest chooses the rating that provides the most votes.

5.12.3.6 K-Nearest Neighbor

KNN is an anti-parametric technique used for correlation or ranking. It is a fundamental methodology that is easily implemented, and it is affordable to construct a design. It includes a data collection where variables are clustered into various groups, and the methodology is intended to recognize the specified test piece of data as a categorization issue. The reliability of the KNN identification is the type of an entity defined by a common vote of its closest neighboring K boors. The categorization of the KNN

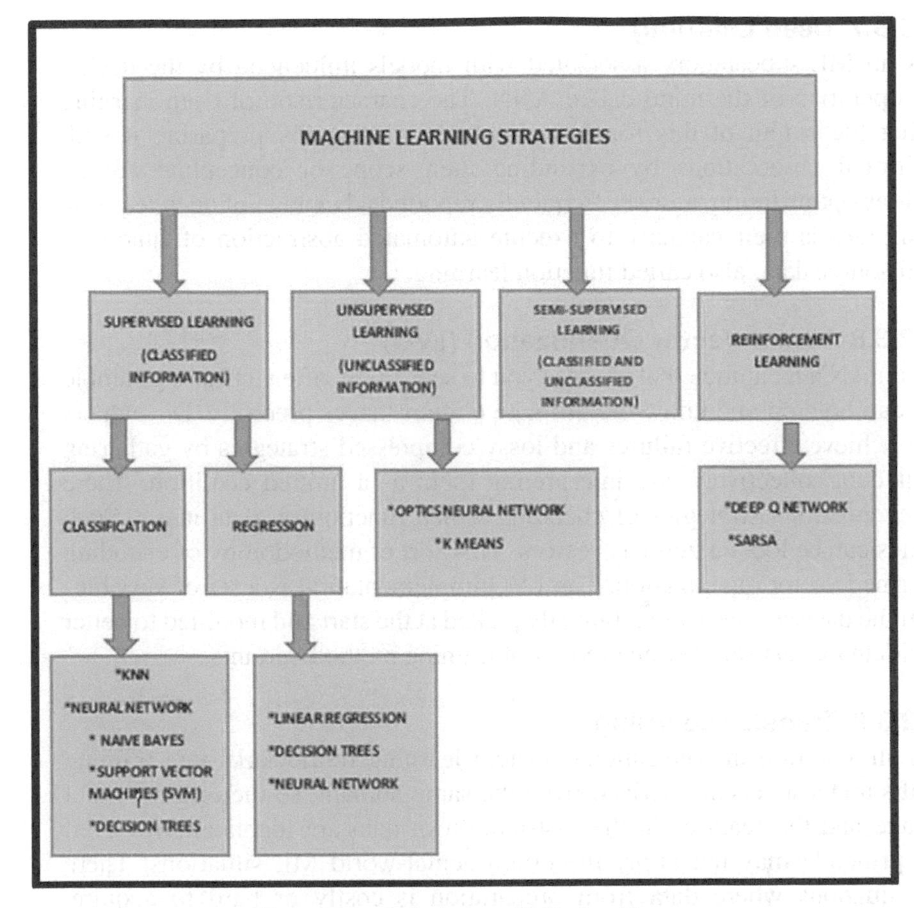

Figure 5.4 Articulation of ML algorithms.

reliability is an entity that is identified by a common vote of its K-Nearest Neighbor. This methodology has the benefits of providing a very comprehensive categorization structure and is highly equipped to cross-modal divisions.

Documents contain several group names; since the failure rate in Bayes will be double that of the average rate, this could be the correct strategy at times. KNN over-performed SVM in predicting features utilizing expression patterns. Its drawback is that KNNs require range estimation. The unidentified documents that need to be listed are broad with the amount established for growth development. Noisy/insignificant characteristics may lead to degradation of consistency.

5.12.3.7 Deep Learning

It is an ML subcategory associated with models influenced by the design and operation of the mind called ANN. The characteristic of deep learning is that the output of this form of structure increases by preparing it with additional observations by expanding their scope or conceptual ability. Besides optimization, another frequently reported advantage of deep learning techniques is their capacity to execute automated abstraction of functions from source data, also called function learning.

5.12.3.8 Linear Vector Quantization (LVQ)

It is an ANN technique that enables you to select how often a training sample you can hold on and know the ultimate circumstances precisely. Developers may achieve effective failures and lossy compressed strategies by gathering input data collectively and interpreting them as a unified coalition. There are even certain strategies of encoding which function on data items. Such frames can be looked upon as vectors. This sort of methodology of encoding is termed vector quantization. The LVQ implementation is a set of variables from the dataset. These are arbitrarily picked at the start and modified to better elucidate the test set over an amount of training method variants.

5.12.3.9 Transfer Learning

It is an assertion in conventional system learning frameworks that training results and test sets are derived from the same domain, so the original input storage and the features in the distribution of data are identical. However, this principle may not apply in certain actual-world ML situations. There are situations where data from preparation is costly or hard to acquire. Hence, there is a requirement to build high-performance trainers qualified from various realms with more readily available information. This technique is called transfer learning which provides details on existing approaches and evaluates implementations described for transfer learning. The survey transition learning approaches are autonomous of information length and can be implemented in massive data contexts.

5.12.4 Non-Parametric Techniques

A few other characteristics of unsupervised learning models are learned from the results. Use of this tool subsequently developed to illustrate the category of information when adding additional information. It is primarily utilized for the segmentation and function minimization [10].

5.12.4.1 K-Means Clustering

K-means clustering technique is employed to organize entities into K numbers of objects based on characteristics and attributes. K is a consistent integer number. Clustering by lowering the quantity of range squares among information and the associated group of centroids. The purpose of the K-mean segmentation is to identify details.

5.12.4.2 Principal Component Analysis

PCA is a statistical technique that uses a set of observations of possibly selected variables to turn them into a set of discrete values called central components. It is employed as a tool for evaluating the information and construct forecasting analytics. To promote and pace numbers, the PCA, the information component, is minimized.

5.12.4.3 A Priori Algorithm

It is a method used by multiple item-set processing and correlation standards to know about external functional sets of data. It progresses by categorizing the different specific items in the collection and expanding them to broader and more generalized attribute values, as well as those object classes that occur in the collection frequently enough. The different element sets *a priori* has discovered and can be used to uncover connection requirements that illustrate cumulative regularities in the data. This has implications in such fields as the review of business array [5].

5.12.4.4 Reinforcement Learning (RL)

RL is an ML field. It is associated with how computer systems can perform steps to enhance the certain principles of accumulated benefit in a system. Apart from unsupervised and supervised learning, RL is among the three techniques for ML. RL distinguishes from supervised learning in not involving the execution of designated inlet and outlet collections and in not demanding specific inadequate conduct. Instead, the emphasis is on striking an equilibrium among exploring exploitations.

5.12.4.5 Semi-Supervised Learning

It offers opportunity as an analytical method for understanding learning regarding individual classification; much of the input is interpreted by itself. This learning aims to consider the various reductions of training quality and model strategies that reap the benefits of such a mixture by merging categorized and unidentified information. There are two forms of semi-supervised model: peer-training and transductive.

5.13 Machine Learning Technology Assessment Parameters

An assessment of the method of ML is an integral component of a description. The approach gives you consistent results when analyzed using a criterion, say, reliability ranking; though once examined toward different criteria, such as logarithmic failure, or another such parameter, it may yield bad outcomes. Often we seek to use classified precision to carry through our prototype's performance, but it is just not enough to pick our system. We will describe different types of evaluations of parameters in this segment.

5.13.1 Ranking Performance

The accuracy of classification guarantees that the definition is used appropriately. It is the ratio of the exact assumptions to the overall amount of input components. Percentage of relevant observations is an overall number of assertions made. This functions better, even though the number of observations per category is similar.

5.13.2 Loss in Logarithmic Form

Regulations design performance identification where a forecast input is a probable rate around 0 and 1. This loss is calculated by the formula given in Equation (5.3).

$$LogarithmicLoss = \frac{-1}{N}\sum_{i=1}^{N}\sum_{j=1}^{M} y_{ij} * \log(p_{ij}). \tag{5.3}$$

Here, *yij* means sample *i* relates or not to class *j* and *pij* means a test probability that *i* contributes to category *j*. Log loss has no upper limit and arises within 0. Log loss closer to 0 means a higher precision although it implies a smaller accuracy if this loss is farther from 0. Reducing this loss essentially provides additional consistency to the classification.

5.13.3 Assessment Measures

There are three purposes employed to implement the assessment validation, notably efficiency, precision, and recall.

5.13.3.1 Accuracy

Accuracy utilizes to conduct the assessment of the framework. This is the component toward each projected example of any actual expected case, which is the predictor of how much is accurate. It is determined by Equation (5.4).

$$Accuracy = \frac{TP + TN}{TP + TN + FP + FN}. \tag{5.4}$$

5.13.3.2 Precision/Specificity

Precision employs for evaluating the reliability of the analysis, in which all favorable feedback is correctly classified as positive based on the amount of comments labeled as positive altogether. It is determined as per Equation (5.5).

$$Precision = \frac{TP}{TP+FP}. \tag{5.5}$$

5.13.3.3 Recall

It is used to calculate correctness. It provides the number of appropriately graded feedback in the check set out from the overall amount of comments as favorable, which are genuinely favorable. The recall is evaluated by Equation (5.6).

$$Recall = \frac{TP}{TP+FN}. \tag{5.6}$$

5.13.3.4 F-Measure

In this retrieval metrics and accuracy criterion. This incorporates the principles of the retrieval with the indicators of reliability. In F-Measure the retrieval metrics and accuracy criteria incorporates the principles of the retrieval with the indicators of reliability. If F-measurement is elevated, then the design of the system is acceptable, and the recommended methods are successful. F-dimensions are determined by Equation (5.7).

$$F - measure = 2 * \left(\frac{Precision * Recall}{Precision + Rcall}\right). \quad (5.7)$$

5.13.4 Mean Definite Error (MAE)

The MAE is the cumulative deviation among the expected results and the observed. Utilized to estimate how much further from the overall result the forecasts were, yet, they do not give an insight into the course of the mistake, that is, if we forecast the data. It is expressed as in Equation (5.8).

$$Mean\ AbsoluteError = \frac{1}{N}\sum_{j=1}^{N} |y_j - \hat{y}_j|. \quad (5.8)$$

5.13.5 Mean Quadruple Error (MSE)

MSE requires a standard square of the variation among the actual results and the expected results. Its advantage is that it is possible to assess gradients, whereas mean absolute error involves sophisticated linear analysis techniques to estimate the gradient. This is as described in Equation (5.9).

$$MeanSquaredError = \frac{1}{N}\sum_{j=1}^{N} (y_j - \hat{y}_j)^2. \quad (5.9)$$

5.14 Correlation of Outcomes of ML Algorithms

Experts have dedicated to ML, and a multitude of methods and strategies have been developed, while five-key optimization learning is part of this methodology. This study utilizes five algorithms, which are NB, SVM, DT, KNN, and RF. Several authors have used the five-algorithm principle to further enhance efficiency, either specifically or with optimization for an upgrade. The segment above describes those five equations.

Presently, to forecast diabetes, the efficiency of such simulations was loosely correlated using the data. Its components involve 786 occurrences and 8 characteristics/qualities. These are standardized utilizing Python source code and are classified as such. All of those implementations were subjected to similar content and similar function collection method. The architectures were evaluated based on the expected period for reliability, $F1$-measure, consistency, and preparation, focusing on the results obtained that demonstrates Classifier's excellent outcomes by using precision calculations. However, the quality of optimization, in particular, relies on the context and the collection of data to which it is implemented. Within diverse circumstances, these methodologies will surpass the other in productivity.

5.15 Applications

The technique of ML has been certified by entities, which cope with massive amounts of data, organizations around numerous fields. Several realms are included in the development of this learning.

5.15.1 Economical Facilities

Organizations can recognize crucial details of specific information in the monetary services global market and even validate instances of money laundering across ML. This also incorporates automated learning methodologies to classify instances of sales and trading. Implementing virtual surveillance can enable individuals or firms to figure out who is prone to economic threats and take proactive measures to mitigate infringement.

5.15.2 Business and Endorsement

In advertising and promotion, several businesses are presently employing machine analytics mechanisms to examine consumer behavior with their clients and offer an individualized brand facility toward their upcoming transaction. These strategies in the long term of distribution and branding could recognize, interpret, and utilize client information to construct a tailored shopping environment.

5.15.3 Government Bodies

They possess a significant requirement for machine analytics in government entities, such as infrastructure and regulatory safeguards; they provide numerous information sets that might be exploited to recognize efficient methods and perspectives. For instance, sensor information can be examined to classify approaches to reduce expenses and enhancing reliability. Implementing ML could violate identification and reveal fraud.

5.15.4 Hygiene

System learning is a crucial increasing medical field with the emergence of wearable systems and detectors that utilize information to reveal patient well-being in the actual moment. Sensors provide patient information in the actual period, like a general medical condition, heart rate, heartbeat, and critical specifications. These details can be interpreted systematically to evaluate the health status of anyone by building a pattern from the history of the individual and predicting the likely existence of any illnesses. The platform also enables medical practitioners to assess data and identify trends that can eventually render effective decisions and treatments.

5.15.5 Transport

It is centered on tradition and trends of traveling through various pathways. These strategies allow the transit corporation to anticipate potential issues with the path and hence recommend their clients to pick an alternative way. Transit probably incorporates this method to identify choice – creating information and model details – and to assist users to make choices throughout their trip.

5.15.6 Fuel and Energy

Immediately by investigating deep reserves and discovering additional forms of resources to channel oil supply. Exclusively by investigating deep reserves and discovering of additional forms of resources to channel oil supply, the requirements for ML are massive and still rising in these sectors.

5.15.7 Spoken Validation

The existing voice identification systems employ these algorithms within the sector to guide the framework for greater precision. Almost all programs incorporate two learning stages in exercise: before the shipping, audio-independent training and audio-dependent mentoring after the shipping.

5.15.8 Perception of the Device

Most current recognition programs, say fingerprint recognition systems, process simultaneously categorizing microscopic images of cells and use higher precision ML techniques, e.g., the North American country post-working environment uses a computer recognition application with a writing detector so that they are eligible to classify letters with handwritten descriptions with a precision rate automated.

5.15.9 Bio-Surveillance

Numerous states desire utilizing ML techniques to track possible communicable diseases. The report collects details on medical treatment department enrolment. The device learning firmware is configured to distinguish deviant indications, their patterns, and the amount of isles employing approved person records. The task is underway to implement certain additional information into the framework, including monitor medications purchasing records and involve further datasets. Just deep learning models can effectively handle the magnitude of flexible and intricate sets of data of such a sort.

5.15.10 Mechanization or Realigning

ML tools are usually used in autonomous systems and machinery. These methods are used by a significant number of data-excessive scientific fields in much of its analysis. For instance, it is often used to classify particular celestial bodies in biology, neuroscience, and psychiatry and behavioral science.

5.15.11 Mining Text

In ML, information extraction relates to the usage of these learning tools to assess data. The mechanism of text processing includes the collection of data and dictionary evaluation to extract data from the document. Standard text analytics involves the classification of data and the convergence of content. The implementations of text mining are safety implementations, perception assessment, internet media functionalities, corporation, advertising technologies, etc. Data analysis strategies are anticipated to convert the document into statistical information, using the natural language processing (NLP) and analytical methodologies.

5.16 Conclusion

This chapter addressed a comprehensive analysis of the automated learning issues in the sense of data analytics. Utilizing massive data descriptions to classify the system learning complexities allows for the development of source impact correlations for each of the challenges. It satisfies this job's initial target of providing a base for a clearer understanding of data analysis computer vision. Another focus of this research was to supply investigators with a solid framework for making easier and better-informed choices about Big Data ML. This document facilitates the implementation of connections in this area of study between the numerous issues and concerns that were not easily doable based on current documentation. Following conventional programming methodologics, managing the dynamic Big Data became complicated with the advances of huge information development. Furthermore, it involves numerous sophisticated, powerful, and logical supervised learning to tackle the massive swathes of classification models. The evidence gathered from these empirical approaches offers increasingly accurate outcomes to several actual-world concerns in diverse aspects such as health, farming, media platforms, finance, etc. Several scholarly reports have been reviewed for accumulating insights on specialized learning strategies. This document provides a broad overview of the sophisticated software learning strategies and processes employed to offer remedies to the challenges of major information optimization. Consequently, this report has fulfilled its ultimate goal by offering possible pathways for subsequent study for the intellectual sector and will ultimately act as foundations for significant changes in the area of deep training with large datasets.

References

[1] Lin Li, S. Bagheri, H. Goote, A. Hasan, and G. Hazard, "Risk adjustment of patient expenditures: A big data analytics approach," in 2013 IEEE International Conference on Big Data, 2013.

[2] G. Zhang, S.-X. Ou, Y.-H. Huang and C-R. Wang, "Semi-supervised learning methods for large scale healthcare data analysis," International Journal of Computers in Healthcare, vol. 2, pp. 98-110, 06/01/2015 2015.

[3] J. Suzuki, H. Isozaki, and M. Nagata, "Learning condensed feature representations from large unsupervised data sets for supervised learning," in Proceedings of the 49th Annual Meeting of the Association for Computational Linguistics: Human Language Technologies: short papers - Volume 2, 2011, pp. 636- 641.

[4] A survey of transfer learning Karl Weiss, Taghi M. Khoshgoftaar and DingDing Wang Weiss et al. J Big Data (2016) 3:9 DOI 10.1186/s40537-016-0043-6

[5] K. Sree Divya, P.Bhargavi and S. Jyothi, "Machine Learning Algorithms in Big data Analytics," International Journal of Computer Sciences and Engineering, vol. 6, Issue-1, E-ISSN: 2347-2693, 2018.

[6] Alexandra L'Heureux, Katarina Grolinger, Hany F. ElYamany and Miriam A. M. Capretz, "Machine Learning with Big Data: Challenges and Approaches," IEEE 2017, DOI: 10.1109/ACCESS.2017.2696365, ISSN: 2169-3536, 2017.

[7] Athmaja S., Hanumanthappa M. and Vasantha Kavitha, "A Survey of Machine Learning Algorithms for Big Data Analytics," International Conference on Innovations in Information, Embedded and Communication Systems (ICIIECS), 2017.

[8] Shweta Mittal and Om Prakash Sangwan, " Big Data Analytics using Machine Learning Techniques," 9th International Conference on Cloud Computing, Data Science & Engineering (Confluence), 978-1-5386-5933-5/19/ 2019, IEEE, 2019.

[9] Imad Sassi, Sara Ouaftouh and Samir Anter, "Adaptation of Classical Machine Learning Algorithms to Big Data Context: Problems and Challenges," 1st International Conference on Smart Systems and Data Science (ICSSD), IEEE Xplore: 20 February 2020, DOI: 10.1109/ICSSD47982.2019.9002857, ISBN: 978-1-7281-4368-2, 2020.

[10] Bilal Abdualgalil and Sajimon Abraham, "Applications of Machine Learning Algorithms and Performance Comparison: A Review," International Conference on Emerging Trends in Information Technology and Engineering (ic-ETITE), IEEE Xplore, 2020.

[11] H. V Jagadish, J. Gehrke, A. Labrinidis, Y. Papakonstantinou, J. M. Patel, R. Ramakrishnan, and C. Shahabi, "Big Data and its Technical Challenges," Communications of the ACM, vol. 57, no. 7, pp. 86–94, 2014.

[12] J. Qiu, Q. Wu, G. Ding, Y. Xu, and S. Feng, "A Survey of Machine Learning for Big Data Processing," EURASIP Journal on Advances in Signal Processing, vol. 67, pp. 1–16, 2016.

[13] J. L. Berral-Garcia, "A quick view on current techniques and machine learning algorithms for big data analytics", 18th International Conf. on Transparent Optical Networks, pp.1-4, 2016. DOI: 10.1109/ICTON. 2016.7550517.

[14] M. U. Bokhari, M. Zeyauddin and M. A. Siddiqui, “An effective model for big data analytics”, 3rd International Conference on Computing for Sustainable Global Development, pp. 3980-3982, 2016.
[15] P. Y. Wu, C. W. Cheng, C. D. Kaddi, J. Venugopalan, R. Hoffman and M. D. Wang, “–Omic and Electronic Health Record Big Data Analytics for Precision Medicine”, IEEE Transactions on Biomedical Engineering, vol. 64, issue 2, pp. 263-273, 2017. DOI: 10.1109/TBME.2016.2573285.
[16] M. R. Bendre, R. C. Thool and V. R. Thool, “Big data in precision agriculture: Weather forecasting for future farming”, 1st International Conf. on Next Generation Computing Technologies, pp. 744-750, 2015. DOI:10.1109/NGCT.2015.7375220.

6

Resource Allocation Methodologies in Cloud Computing: A Review and Analysis

Pandaba Pradhan[1], **Prafulla Ku. Behera**[2] **and B. N. B. Ray**[3]

[1]*Asstistant Professor & H.O.D, Department of Computer Science*
BJB (Auto) College, BBSR, Odisha, India.
[2]*Reader & H.O.D, Department of Computer Science & Applications.*
Utkal University, BBSR, Odisha, India.
[3]*Reader, Department of Computer Science & Applications.*
Utkal University, BBSR, Odisha, India.
E-mail: [1]ppradhan11@gmail.com, [2]p_behera@hotmail.com,
[3]bnbray@yahoo.com

Abstract

Cloud computing has emerged as a new technology that has got huge potential in enterprises and markets today. Clouds can make it possible to access applications and associated data from anywhere. Companies are able to rent resources from cloud for storage and other computational purposes so that their infrastructure cost can be reduced significantly. Cloud technology is an on-demand infrastructure platform that provides pay-per-use services to cloud users. In order to efficiently allocate resources and to meet the needs of users, an effective and flexible allocation of resources mechanism is required. According to the growing requirements of the customers, the resource allocation process has been more complex and complicated. One of the main priorities of research scholars is on how to design effective solutions for this problem. A literature review on current competitive resource allocation methods is included here.

Keywords: Cloud, cloud computing, resource allocation (RA), task scheduling, resource allocation strategy (RAS), quality of service (QoS), service-level agreement (SLA).

6.1 Introduction

Resource allocation (RA) is a big challenge of cloud computing as well as any other computing environment. Cloud computing helps to provide quick and efficient access to computing infrastructure including servers, networks, storage, and likely facilities. Cloud vendors must maintain, deliver, and distribute these tools effectively and offer cloud-based services to customers on the basis of service-level agreements (SLAs) that the parties agree before the customer uses the services. Providers must also ensure a robust allocation process to meet the needs of cloud customers, at almost the same period; they stabilize an acceptable profit margin for themselves.

The static allocation and resource management have become inefficient in cloud environments, and the implementation of dynamic methods has become more relevant and worth researching. However, also these complex processes bring difficulties and obstacles to be resolved and solutions to be found. Many researchers have attempted, and are still working, to provide effective solutions for resource utilization and management problems in cloud computing environments.

6.1.1 Cloud Services Models

Cloud computing is a web-based technology which provides users with infrastructure as a service (IaaS), platform as a service (PaaS), and software as services (SaaS). SaaS is a cloud service design that is conducted on the server and provides cloud applications to consumers that run on its infrastructure. Customers had no influence over service management. Examples of SaaS applications are: content management systems, customer relationship management assistances, video conferencing, and e-mail communication systems. Cloud customers in PaaS have the capability to build and run software applications using the cloud provider-supported programming languages, facilities, databases, and resources. The user does not have the capability to handle or monitor the deployed network, yet has influence over the software being installed and may customize the settings. IaaS is a cloud platform that provides customers/users with certain basic computing facilities such as processing power, memory, networks, and other computer resources on demand to deploy and run their applications on them. Customers may only run their program on the cloud services, which do not have power over network management. In fact, there are more service models in use today than the three (IaaS, PaaS, and SaaS), such as system models such as data analytics as a service and HPC/grid as a service appear as practical

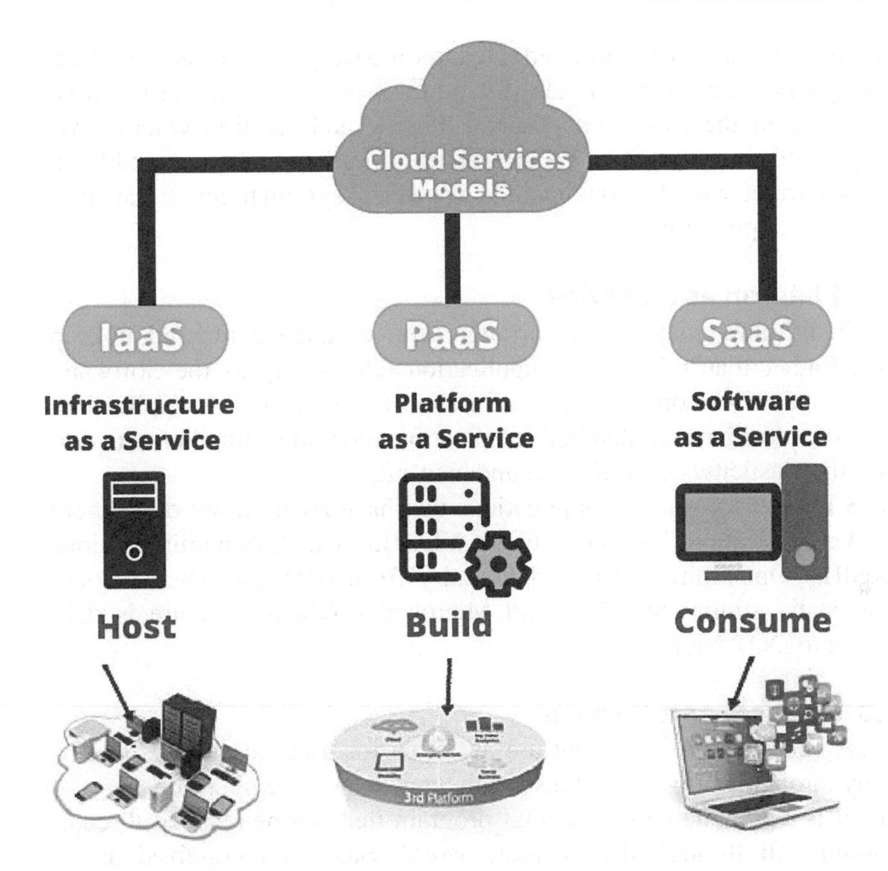

Figure 6.1 Cloud computing services model.

models. How to choose a suitable business model depends on factors such as availability of appropriate applications, the prerequisite for a production and testing environment, the need for efficient regulation of the network, management of the necessary delivery of data, resources and facilities, and sophistication of the enterprise IT technology and data center/warehouse. The three main models of cloud services are listed in Figure 6.1.

6.1.1.1 Infrastructure as a Service

IaaS is generally utilized for web facilitating, in which a web worker and a working framework stack are based on virtual machines (VMs), where cloud administrations, for example, quick scaling, local versatility, overseen setting, geographic burden adjusting, uncommon front-end content conveyance, or availability can normally be utilized. Disaster recovery prevention, for which a running server snapshot is made (including in a memory image), can be reconstructed in the case of a disaster. Testing and creation where VMs are easy to fabricate ought to be utilized to drop the need to assemble an exploration framework that may include a powerful system to investigate new advances under pressure.

6.1.1.2 Platform as a Service

Here, clients have control over arrangement, establishment, and upkeep of uses. Rather than utilizing an application relationship as the Software, the executives and conveyance measure abstracts from the application. Service providers have all detailed information about the computer network, availability, elasticity/scalability, redundancy, etc.

PaaS is ideal for running applications that have risen up out of the field of “holder applications” before J2EE or NET. In fact, PaaS administrations like RedHat OpenShift and Cloud Foundry from IBM, and even Pivotal, are practically similar to J2EE, and Microsoft’s Windows Azure is only equivalent to DOT NET.

6.1.1.3 Software as a Service

There are full software apps that are available for users to access directly, normally through a browser, but also with any “done” user interface.

Email is a genuine case of a total program that can be reached through a program. All through the versatile world, SaaS is recognized as an application, while the front-end user interface (UI) is on the telephone, when the back end is in the cloud. In actuality, though other mobile applications do not resemble that, they are worked with the adaptable versatile adaptation of a Web Kit program.

SaaS sellers should likewise ensure a sensible norm of assurance by acquiring the particular affirmation and consistence level required in that industry and afterward by guaranteeing additionally that hidden cloud specialist co-op keeps up the particular confirmation and freedom level required in that condition.

6.1.2 Types of Cloud Computing

In cloud computing technology, vast quantities of computer resources are linked for storage and exchange of knowledge. This can be done via public, private, mutual, and hybrid cloud as illustrated in Figure 6.2.

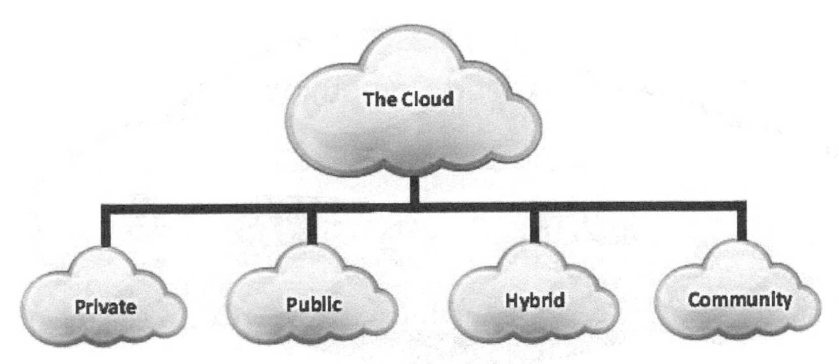

Figure 6.2 Types of cloud computing.

6.1.2.1 Public Cloud

The NIST idea is: Distributed computing is intended with the expectation of complimentary access by the overall population. It very well may be claimed and overseen by, or a mix of, a partnership, scholastic, or government substance.

It shows up at the cloud supplier premises and is a type of giving open cloud administrations and the cloud office supplier plan of action bringing economies of the size of server farm adjusting, virtualization just as on-request coordinated replication empowers business IT foundation re-appropriating takes care of the debacle recovery issue fitting for small and medium enterprises (SMEs) and nimble organizations.

Nevertheless, the data and transmission system are not under the jurisdiction of organizations and may pose questions regarding software or data with a security need.

These are distributed computing assets that are made accessible by open cloud specialist organizations that oversee and work the foundation and are available to the doorsteps of the wide range of the nation. Open cloud might be possessed, worked, and constrained by a blend of scholarly establishments, business associations, or government associations. Instances of overall population cloud administrations comprise online stockpiling administrations, email administrations, person-to-person communication sites, and applications, for example, Facebook, WhatsApp, and Viber. The description of the public cloud vendors are displayed in Figure 6.3.

6.1.2.2 Private Cloud

The NIST concept is: A single entity containing several users (e.g., corporate units) supplies the cloud services with exclusive use. This may be owned,

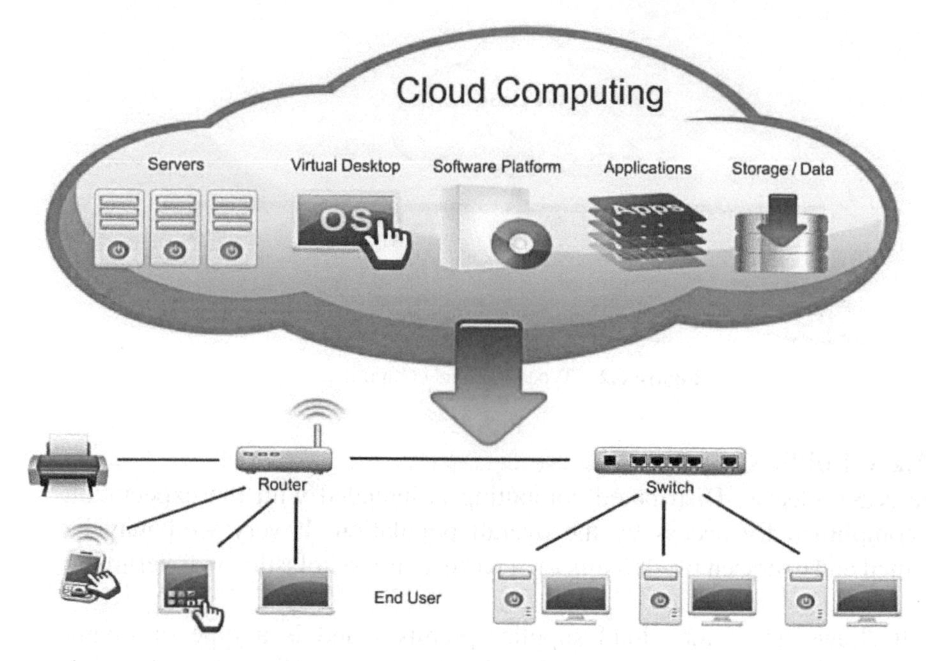

Figure 6.3 Summary of public cloud providers.

maintained, and run by, or a combination of, the company, a third party provider, and it could occur on bad locations.

It is a cloud computing platform that provides IT to consumers at a reduced cost as a commodity, rather than a product. It provides higher productivity to consumers with enhanced, creative business prototypes that are appealing and recommended for their business growth and transformation in this digital age. It is owned and controlled by private companies, third parties, or both combined.

Private clouds are an option for organizations that already own data centers and build the infrastructure that have different security or efficiency requirements. They are really a better option for the company data center than integrated services in several ways, trying to bring many potential benefits from virtualization and automation.

6.1.2.3 Community Cloud

Network cloud is a multi-seller framework that falls between both general society and private cloud as for their market part, offering cloud administrations to some network of individuals or customers inside a substance with comparable premiums or determinations (e.g., target, security, and protection strategy and implementation contemplations). The expense of securing the cloud administrations in the gathering cloud is less exorbitant when contrasted with open and private cloud, as it is disseminated among associations.

This may be owned, maintained, and controlled by individual or extra voluntary groups, a third party or a combination thereof, and can be present on off-premises. The Group Cloud would also include coordination and convergence of IT systems and services from various organizations. It can serve large inter-organizational projects.

6.1.2.4 Hybrid Cloud

It is a distributed computing office that utilizes a blend of two or probably more unique cloud stage (open, private, or network) with explicit substances which permit the sharing of remaining tasks at hand between both the two cloud specialist organizations to empower information and convenience of utilizations. For instance, cloud blasting to adjust load between mists.

Mixture cloud, while the most intricate design to oversee, is additionally the most practical model for current organizations. It consolidates center cloud-based venture framework with high-load redistributing errands close to open mists. It comparatively partners the upsides of a controlled domain of private mists and the expedient flexibility of open mists. In any case, it requires a more profound IT/cloud modernization of the endeavor. Cycles/work processes require reproducing.

The problems of smooth convergence among the private and the public cloud which must be overcome by tailored research, and accessibility, standardization and security concerns are not yet key issues.

6.2 Resource Allocations in Cloud Computing

In distributed computing, asset allotment is an idea that is thought about in numerous regions of processing, for instance, server farm activities, working frameworks, and lattice registering. Asset allocation includes the assignment of accessible assets between cloud clients and projects in a monetary and gainful way. It is novel from the most troublesome exercises of IaaS-based distributed computing. Furthermore, RA for IaaS in distributed computing takes a few focal points: it is financially savvy since the buyer does not need to introduce and update equipment or programming to get to administrations, its flexibility empowers admittance to projects and information on any gadget on the planet, and there are no limitations on the stage or spot of utilization. There are two primary RA components in distributed computing.

6.2.1 Static Allocation

Static allocation scheme allocates a specific service to a cloud customer or to a program. In this situation, the cloud customer should be mindful of the amount of resource instances required for the program and the services provided and should attempt to validate the peak load requirements for the application. Be that as it may, the impediment of static assignment is generally influenced by over-use or under-utilization of registering assets relying on the ordinary outstanding task at hand of the application. This is not generally financially savvy and is because of an absence of use of the asset during the off-top time frames.

6.2.2 Dynamic Allocation

Dynamic allocation scheme offers cloud assets on the spot whenever a cloud client or program is required, essentially to forestall over-use and under-utilization of assets. A potential drawback when required administrations are requested on the spot is that they may not be accessible. The specialist co-op will, at that point, disseminate staff from the different taking an interest cloud information centers. The resource allocation strategy (RAS) identifies with the mix of cloud office supplier tasks for the utilization and arrangement of accessible assets inside the cloud foundation to fulfill the necessities of the cloud application.

As cloud computing has its own characteristic features, the RAS must try to prevent the subsequent conditions as far as possible:

Resource contention: This condition happens when multiple users and programs attempt to use the same data concurrently.

Resource fragmentation: This condition arises when applications are unable to allocate resources based on the fact that isolated resources are small items.

Resource scarcity: This condition happens when various resource demands for applications are high and there are insufficient resources, such as memory needs, I/O devices, CPUs, and procedures which can remain used to satisfy this need.

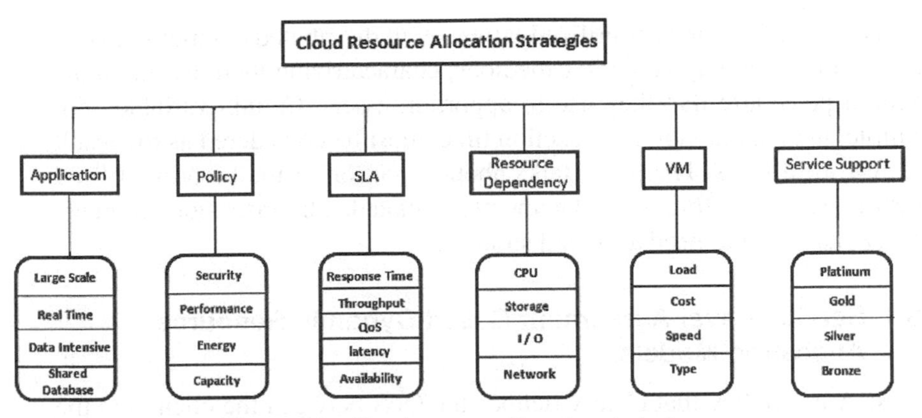

Figure 6.4 Cloud resource allocation strategies.

Resource over provisioning: This condition happens when customers and programs receive more resources than any of those needed to fulfill the quality of service (QoS) standards.

Resource under provisioning: This condition happens when users and programs are provided with less resource than any of those required to meet with the QoS standards.

From the perspective of cloud clients, RA must be figured out how to accomplish at a lesser expense and with as meager time as could be expected under the circumstances. Notwithstanding, it is not workable for the cloud luxury suppliers to gauge the intricacy of client necessities, the idea of clients, and, furthermore, the application prerequisites. Accordingly, asset multi-faceted nature, insignificant space, nearby constraints, the mind boggling type of information requests, and ecological issues need a viable and helpful RAS that is suitable for cloud situations.

As the dynamics and unknown risks of resource demand and supply are inconsistent, a variety of dynamic RA strategies are recommended. This work discusses different RASs that have been found in cloud environments.

6.3 Dynamic Resource Allocation Models in Cloud Computing

The quality and cost of distributed computing administrations depend on their RA cycle, just as the asset supplier ought to pleasantly relegate the asset to clients. Be that as it may, there are a few RA techniques and proposed models which are utilized famously in distributed computing. Here are a portion of the dynamic RA methods, characterizing them based on the essential procedure that they use to apportion assets. Certain variables, for example, inertia, amount, and reaction time, must be considered as the result of a few ideal RAS. The paper talks about a portion of the broadly utilized procedures: administration level game plan situated, administration arranged, market situated, and need arranged strategies.

6.3.1 Service-Level Agreement Based Dynamic Resource Allocation Models

The SLA is a policy that clearly defines the QoS between the client and the facility provider also provides the QoS-level service costs defined mostly by quality of the service. The bulk of RA methods in cloud services usually depends on maintaining the specified SLA standards regarding cloud users. A lot of other RA models rely on meeting the goals of the cloud service provider, which may adversely impact some of the user's criteria and the type of QoS offered.

However, one concept is suggested by I. Popovici and J. Wiles. They explored the QoS parameters, like the load given and the cost on the SaaS provider's part, but did not reflect the customer's aspect [1].

S. Farokhi has developed a multi-cloud resource management system from the SaaS-level perspective, approved SLA, as well as facility provider terms and conditions. The suggested model uses the selection engine, construction system, and SLA intrusion identification and tracking in the use of QoS criteria by the service provider [2].

There are barely any models that focus on both the cloud stage and the client. Such a thought has been started by Wu, Linlin, Saurabh Kumar Garg, and Rajkumar Buyya, which focused generally on QoS prerequisites of both SaaS suppliers. Just as the purchaser by presented RA calculations planned for lessening administration level arrangement (SLA) penetrates and framework costs, and furthermore checking explicit client alteration, through deciding client prerequisites for foundation level part and the board of the disparity of VMs. The two proposed calculations execute well by lessening costs by practically half with diminished VMs and altering the way to forestall pointless help level understanding (SLA) harms [3]. In examination, the other exploration introduced by Lee Gunho *et al.* centers around the issue of advantage, generally on premise of administration necessity arranging in distributed computing, considering the needs of the two clients and specialist co-ops [4].

Zhu *et al.* suggested structure to resolve virtualized RA challenges with multi-level applications. Their prototype increased product efficiency, reduced costs, maximized profitability for SaaS providers, and was designed to satisfy consumer performance criteria.

Efficient agent based resources allocation (EARA) is a reliable agent-based RA system developed by Kumar, Ajit, Emmanuel S. Pilli, and R. C. Joshi. EARA used the agent computing as a number of agents acquire available resource details for user requests on the basis of the SLA agreement signed and therefore balance performance and cost control [5].

Saurabh Kumar Garg and Rajkumar Buyya [7] likewise actualized a market-arranged RA technique by fusing every client-driven vital arranging and administration-driven danger appraisal to energize use of assets depending on the administration level understanding (SLA). Notwithstanding, it needs a market maker and a market library to hold together clients and sellers, to distribute cloud assets and to find their merchants [7].

What is more, Pawar, Chandrashekhar S., and Rajnikant B. Wagh proposed a need-based allotment model by considering diverse assistance level arrangement (SLA) standards, for example, data transfer capacity, force, and season of execution. Utilizing an acquisition system related with the upside of equal preparing, their model improved proficiency, particularly in the feeling of asset dispute [8].

6.3.2 Market-Based Dynamic Resource Allocation Models

The utilization of the commercial center economy to deal with RA has been generally explored previously, and numerous researchers have inspected the financial ramifications of distributed computing from different points of view. Market-based RA of the executives has been acquainted with tackle, quickly fluctuating asset requests and has been embraced by an assortment of open IaaS sellers, for example, Amazon EC2 [12]. Under this situation, the cloud office supplier may follow a wares market methodology or sale-based methodologies, with the essential target of producing maximal gainfulness while decreasing expenses.

Zaman S. proposed a mixture selloff-based model for cloud asset the executives. Their calculation depends on the client's valuation standard, specifically that every client requires a specific arrangement of VM occurrences and offers on it. This reflects compelling appropriation and high salary for the provider, which likewise expects customers to pay a fixed least sum [9].

There are many other market-based ideas that justify the equilibrium concept and use a genetic algorithm (GA) based budget-adjusted algorithm, much like the one advised by You, Xindong *et al.*, the RAS-M model. Their model is powerful and increases efficiency and profits; however, it only addresses the base level of the cloud domain and is restricted to CPU properties only [12].

Lin, Wei-Yu, Guan-Yu Lin, and Hung-Yu Wei introduced a cloud RA model based on a locked bid auction where consumers send their bids to cloud service providers receiving bids and deciding the cost. This approach allows for effective management of resources, but then according to its facts-saying rights, no benefit maximization is ensured [13].

Zhang, Qi, Quanyan Zhu, and Raouf Boutaba implemented a spot market model that tackled the problem of allocation of resources for various VMs styles in Amazon spot examples with the model of predictive controller (MPC) algorithm. The suggested framework guarantees stable profits for suppliers over time by adjusting pricing based on the level of competition, satisfying consumer needs and reducing energy usage; however, current and future markets are not included in the framework [10]. From here then, Fujiwara, Ikki, Kento Aida, and Isao Ono enhanced the market-based allocation by implementing a dual-sided hybrid auction-based model that helps both consumers and providers to transfer their existing and potential on-site and forward-looking resources [11].

6.3.3 Utilization-Based Dynamic Resource Allocation Models

So as to determine the under-utilization of assets emerging from the portion of fixed assets to projects and offices, the fundamental methodology of the strategy utilized in this segment is to effectively control VMs to improve asset use and limit costs. Lin, Weiwei, and Deyu have fabricated a model that changes the VMs as per the particular needs of the program and it depends on the edge. The proposed strategy utilizes following and anticipating the requests of cloud applications, which prompts improving asset utilization and lower costs [14].

Yin *et al.* accentuate on RA at the phase of the program. The specialists proposed a multi-dimensional RA (MDRA) structure utilizing application allotment system to decrease the expenses of the server farm by designating restricted hubs to the client program models [15].

Simulated analysis powered RA has been implemented by D. Pandit, S. Chattopadhyay, M. Chattopadhyay, and N. Chaki using a multi-parameter bin packing algorithm to minimize the unallocated portion of the resource variables. The new model increased the usage of cloud resources on a multi-level cloud network and cost savings [16].

Lee, Gunho *et al.* has implemented the topology aware based resource management allocation (TARA) scheme. This scheme tackles the unconcerned demands of the hosting program for the IaaS method. They suggested a forecasting engine and a GA-based search for reduced latency and proper trust. The authors have shown that the TARA experiment could result in a decrease of up to 59% in the time of job compilation [4].

Li and Qiu proposed a cloud adaptive RA calculation for a preemptive employment arrangement. The creators have distinguished two calculations, versatile rundown planning (ALS) and versatile min-min booking (AMMS), which identify with booking undertakings, and have demonstrated that the proposed technique is successful and effective for asset use. Their model builds by and large use by decreasing over-burden execution and vitality utilization and giving generally effectiveness to server farms in the distributed computing climate [17]. J. Li *et al.* proposed online arrangement for booking preemptive undertakings on IaaS cloud gadget models utilizing a min-min calculation, and the planning cycle depends on contribution on the individual activity. Their methodology has appeared to limit the hour of RA execution and vitality utilization; nonetheless, the intensity of their model is subject to the unwavering quality of input information [19].

Rammohan and Baburaj have developed an RA in a cloud computing model based on interference-aware resource allocation (IARA) technology that delivers optimum energy usage and is feasible for a resource-constrained environment and also supports special hardware [20].

There have been a lot of other concepts in this. Many of these models rely primarily on optimizing the utilization of available resources.

6.3.4 Task Scheduling in Cloud Computing

Task scheduling is a process for the assigning of one or more time periods to one or more assets. In the cloud-based environment, with the scheduling scheme, it is hard to schedule a series of tasks performed by various users on a collection of computational services to reduce the end time of a particular task or makespan of a program. Scheduling is a series of rules that control the sequence of work to be done by a computer program.

Different kinds of scheduling algorithms reside in the parallel computing world; the main advantage of the scheduling algorithm is high-performance computation as well as best system performance. Scheduling controls CPU memory capacity, and effective scheduling management offers optimum power usage.

Assignments are put together by the clients to the Data Center Broker in distributed computing environment. The Data Center Broker is a specialist between cloud clients and sellers and is answerable for organizing VM exercises. Server farm is a virtual asset lodging foundation that incorporates various hosts. The undertakings submitted are set out as per the planning approaches utilized by the Data Center Broker. Dealer manages the cloud overseer legitimately and allots tasks to VMs in the Data Center Host.

Due to various strategies, there are several types of scheduling such as preemptive and non-preemptive scheduling, static and fluid scheduling, immediate and batch scheduling, and centralizing and distributed scheduling. Task scheduling is seen as the process of choosing the best possible resource for the execution of tasks. Task scheduling algorithms aim to lessen the finishing time of tasks and optimize the utilization of resources to satisfy the expectations of the client.

In immediate mode, tasks remain arranged as the computation system arrives, and in batch mode, the first tasks are clustered into a batch; that is, a number of meta-tasks will be distributed at times, called mapping cases.

In static scheduling, the tasks are which were before-schedule, all information is known about the resources and tasks accessible in advanced, and also the task is assigned to the resource once.

In dynamic scheduling, tasks are dynamically accessible for managing the scheduler over time. It is much more flexible than static scheduling to be able to evaluate run time in advance.

In preemptive scheduling, each task can be disrupted during implementation and the task can be transferred to some other resource by eliminating the initially assigned available resource for other tasks. If restrictions such as urgency are considered, this scheduling scheme is more effective.

In the non-preemptive scheduling scheme, resources are not allowed to be re-allocated until the preplanned and planned work is completed.

Non-preemptive scheduling has been used while the task is over or where the task is shifted between running to waiting. In this scheduling, once the resources have been assigned to the route, the process will hold the CPU until

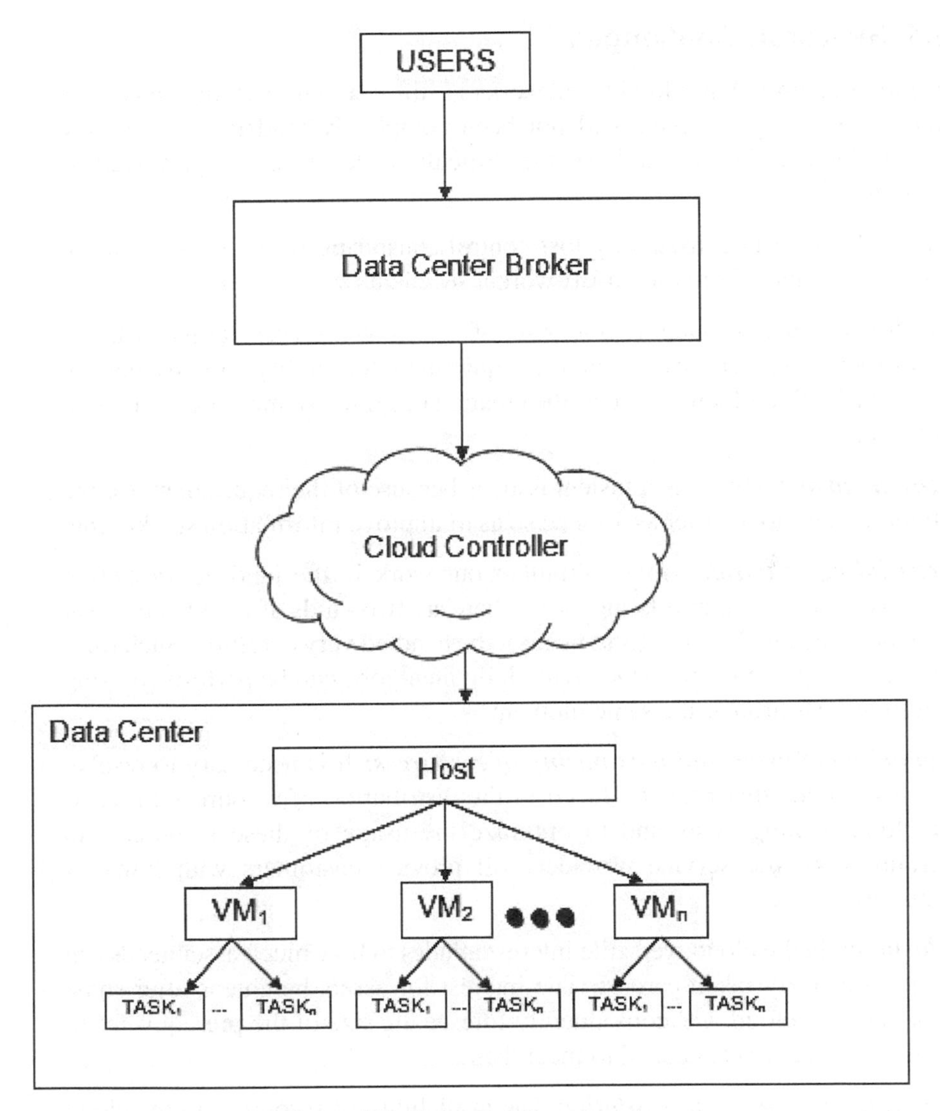

Figure 6.5 Task scheduling structure.

it is terminated or the waiting state is reached. In the case of non-preventive scheduling, the process running the processor should not be interrupted in the middle of the operation. Instead, it waits for the machine to finish its clock burst time, and then the clock can be transferred to another operation.

6.4 Research Challenges

Exploration on RA in cloud frameworks is still in the underlying stages. So many existing issues have still not been completely tended to while new difficulties remain. A couple of the difficult work issues are portrayed as follows:

Control: There is consistently lost control conspiring over the assets when they are obtained from the far-off worker by clients.

Vitality efficiency: Due to the rise of enormous server farms utilizing assortment of server farms, there is a requirement for vitality effective portion of PC tasks. Such focuses add to the creation of immense measures of carbon dioxide.

Movement of VMs: This transient issue is because of the requirement for the client to move to another assistance so as to improve information stockpiling.

Scheduling of parallel jobs: Simultaneous work in the field of computing increases the work that is being used. There are two kinds of work: dependent and independent. The first form needs to be handled very carefully. Such roles require the problems of connectivity. Individual jobs can be performed using multiple VMs around the same moment.

Reduction of price and maximizing of resources: It is necessary to resolve the restrictions that must be faced in the distribution of resources in terms of cloud running costs and to optimize the usage of these resources. In certain ways, the service provider will provide customers with low-cost facilities.

Elasticity: In the cloud, versatile interest alludes to how much asset needs can be progressively taken care of. The interest for assets heightens after some time, and the cloud will consequently foresee the size of the prerequisites to be met and the assets needed to meet them.

Maintaining higher accessibility: The availability of resources in the cloud must be ensured if there is a long-running job that can take many hours to complete. Some techniques are therefore needed to repeatedly handle any disturbance of resources and to switch jobs to the available resource. In addition, these techniques must support the transparency property by which the user is unable to observe a lack of availability or any failed problem.

6.5 Future Research Paths

Many work has been done and a few thoughts have been proposed in the zone of distributed computing concerning the RA issue; besides, there appear to be a couple of issues and hindrances that require more examination, and an ideal methodology that is doable for most cloud conditions has not been found at this point.

A few of the assumptions based on our literature from previous work are as follows:

1. There is a need to reduce customer SLA damages when maximizing RA usage, as most frameworks affect QoS to reduce costs and retain strong efficiency algorithms.
2. An RA scheme is required which is suitable for various cloud environments to solve the utilization issues in various clouds.
3. There is a need for RA to reduce costs for cloud users and increase revenues for cloud providers. This is really important for cloud vendors to make effective usage and maintenance of the finite number of available space.
4. There is a need to understand load balancing of cloud services and to schedule the workload optimally in order to satisfy the users' QoS needs and optimize income by increasing the usage of resources.

The future path of cloud RA research could perhaps address several of the challenges mentioned in the previous sections and continue to implement the best practice models.

6.6 Advantages and Disadvantages

Advantages	Disadvantages
1. The main advantage of RA methodology is that people do not have to install software or equipment to access services, create an application, and manage an application over the web.	1. Even though users rent services for their utilizations from remote servers, they do not have authority over their services.
2. The very next big advantage is that there is no limit of location and medium. You will access your software and data on any device anywhere in the world.	2. The issue of migration arises as consumers decide to migrate to any other service for improved data access. Transferring large data from one provider to the other is not convenient.

3. The customer does not have to spend on software and hardware.	3. Client data can be vulnerable to hacking or phishing attempts in the open cloud. Because cloud servers are inter-linked, the spreading of ransomware is easy.
4. Cloud service providers can distribute their services on the web during resource scarcity.	4. Deeper information is needed for cloud resource distribution and management, as all awareness regarding cloud functioning totally depends on the cloud service provider.

A comparison analysis of different cloud RA frameworks and their advantages and disadvantages are shown in the table below.

Reference number	**Method used**	**Advantages**	**Disadvantages**
[1]	Economics-based approach using utility function	This analyzed the QoS criteria, including the load and price provided by the SaaS provider.	This does not really find infringements of SLA to serve clients.
[2]	Proposed framework assists SaaS providers in multi-cloud system	It helps to identify reliable network resources that better fulfill consumer needs when tracking SLA and identifying infringements.	The complexity and data traffic used in the specified resources in a cloud infrastructure at execution time are not regarded.
[3]	Proposed RA algorithms for SaaS providers	The proposed algorithms increasing the expense of the SaaS provider and the amount of SLA infringements. It handles the diverse integration of consumers.	The net benefit of the algorithms has to be increasing in order to boost the degree of consumer service. It will track the restriction of fines by recognizing device errors.
[4]	TARA schema	It greatly reduces the period for the work to finish the request.	The methods included in this system were never known to be complex objective functions relating to power, operational costs, and efficiency.

[5]	Agent-based algorithm	Allow allocations for multi-level systems. It increases quality efficiency and reduces prices. It eliminates breaches of the SLA for both the customer and the supplier.	Algorithms are only appropriate for the SaaS-type system.
[6]	EARA framework	It helps the consumers with the necessary resources. This is used by a huge variety of people.	If an agent fails, the framework does not perform efficiently.
[7]	Market-oriented mechanism	This offers all client-driven and vendor-driven resource control.	There are other specifications, i.e., market maker and register operator.
[8]	Priority and heuristic scheduling algorithms	This enables the efficient use of cloud assets to meet the SLA objective.	This system cannot estimate VMs that would be freed sooner and rely on its ability to pick a function from waiting for the queue to operate on that VM.
[9]	Combinatorial auction mechanism	This guarantees effective distribution and good income for the vendor. It requires consumers to pay a fixed amount.	Insecure benefits for consumers and customers of utilities cannot share their resources with vendors.
[10]	MPC algorithm	This offers sales enhancement to enable consumer needs to be fulfilled. This minimizes power usage.	Complete absence of expected revenue projection due to lack of understanding for business planning.
[11]	Double-sided combinatorial auction based model	Enable all users and providers to exchange their existing and potential resources on the place and on the future market.	The heavy needs of the system.Complex resource utilization.
[12]	Equilibrium theory and GA-based price adjusted algorithm	This ensures effective management. Earn excess profit for cloud vendors.	This just describes the physical stage of the cloud system and is restricted to CPU resources.
[13]	Sealed-bid auction	It provides for with an efficient utilization of resources.	Unsuitable financial planning due to its facts-telling property.

[14]	A threshold-based dynamic allocation of system resources	This helps increase the usage of resources and lessens prices.	There is no consideration of the expenses of infrastructure.
[15]	MDRA schema	The suggested algorithm enhances the usage of resources and decreases the quality.	It is impractical to save power when the requirement of the consumer rises in a long-termscenario.
[16]	Proposed bin packing algorithm use for simulated annealing	It significantly reduces expenses and deals with issues with the deployment of multi-layer services.	It seems to be a complete absence in managing dynamic resource requests. Compares the availability request and prevents the size of the bin.
[17]	Adaptive list scheduling (ALS) and adaptive min-min scheduling	The strategy maximizes the usage and profit of the cloud service.	Avoids access over victims with SLA offenders.
[18]	A green cloud framework-based scheme	Greatly increases usage. Decreases noise and power usage. Offers complete usefulness for cloud services.	The setup phase of the system is high in cost.
[19]	Min-min and list scheduling algorithm in preemptible tasks	It leads to major changes in the intense resource contention and facilitates load handling. This algorithm has a slightly shorter processing time.Reduces the power usage.	The results could not be correct and also can contribute to over-provisioning or under-provisioning. There is a possibility that response details would be accurate enough as the controller to depend on.
[20]	IARA-based model	It ensures optimum energy usage. It is feasible for a resource-restrictedenvironment and supports special hardware.	Minimum performance and QoS.
[21]	Robust cloud resource provisioning (RCRP) algorithm	The method helps to reduce the overall price of services. This is deployed on the operator side as opposed to each other on the cloud agent line.	It entirely perceives a reserve policy for the provision of resources.

[22]	Proposed RA algorithm	The method significantly increases the cloud provider's net profit. Decreases maintenance costs and infringements of the SLA.	It is requirements of VMs.

6.7 Conclusion

Distributed computing advances are quickly being utilized in enterprises and in the present commercial center. Also, from that point, a successful RAS is expected to accomplish buyer fulfillment and lift benefits for cloud specialist co-ops.

This section portrays an audit of the vast majority of the RA issue structures and arrangements in the distributed computing climate. A few models are sorted based on their methodologies and an investigation of their advantages and downsides is introduced by a correlation table. In conclusion, research ways from our audit of the writing are incorporated and potentially will assist with elevating future scientists to characterize appropriate RA usage for cloud conditions.

References

[1] I. Popovici, And J. Wiles, "Proitable Services In An Uncertain World" In Proceeding Of The 18th Conference On Supercomputing (Sc 2005), Seattle, Wa.

[2] S. Farokhi, "Towards An Sla-Based Service Allocation In Multi-Cloud Environments," Cluster, Cloud And Grid Computing (Ccgrid), 2014 14th Ieee/Acm International Symposium On, Chicago, Il, 2014.

[3] Wu, Linlin, Saurabh Kumar Garg, And Rajkumar Buyya. "Sla-Based Resource Allocation For Software As A Service Provider (Saas) In Cloud Computing Environments." Cluster, Cloud And Grid Computing (Ccgrid), 11th Ieee/Acm International Symposium On. Ieee, 2011.

[4] Lee, Gunho, Et Al. "Topology-Aware Resource Allocation For Data-Intensive Workloads." Proceedings Of The First Acm Asia-Pacific Workshop On Workshop On Systems. Acm, 2010.

[5] Kumar, Ajit, Emmanuel S. Pilli, And R. C. Joshi. "An Efficient Framework For Resource Allocation In Cloud Computing. Computing,

Communications And Networking Technologies (Icccnt), Fourth International Conference On. Ieee, 2013.

[6] Jiao, Jianxin Roger, Xiao You, And Arun Kumar. "An Agent-Based Framework For Collaborative Negotiation In The Global Manufacturing Supply Chain Network.Robotics And Computer-Integrated Manufacturing 2006.

[7] Garg, Saurabh Kumar, And Rajkumar Buyya. "Market-Oriented Resource Management And Scheduling: A Taxonomy And Survey. Cooperative Networking 2011.

[8] Pawar, Chandrashekhar S., And Rajnikant B. Wagh."Priority Based Dynamic Resource Allocation In Cloud Computing With Modified Waiting Queue." Intelligent Systems And Signal Processing (Issp), International Conference On. Ieee, 2013.

[9] Zaman, Safdar; Grosu, Daniel. "A Combinatorial Auction-Based Mechanism For Dynamic Vm Provisioning And Allocation In Clouds" Cloud Computing, Ieee Transactions, 2013.

[10] Zhang, Qi, Quanyan Zhu, And Raouf Boutaba. "Dynamic Resource Allocation For Spot Markets In Cloud Computing Environments" Utility And Cloud Computing (Ucc), Fourth Ieee International Conference On. Ieee, 2011.

[11] Fujiwara, Ikki, Kento Aida, And Isao Ono. "Market-Based Resource Allocation For Distributed Computing" Vol. 2009. Ipsj Sig Technical Report, 2009.

[12] You, Xindong, Et Al. "Ras-M: Resource Allocation Strategy Based On Market Mechanism In Cloud Computing" In: 2009 Fourth Chinagrid Annual Conference. Ieee, 2009.

[13] Lin, Wei-Yu, Guan-Yu Lin, And Hung-Yu Wei. "Dynamic Auction Mechanism For Cloud Resource Allocation." Cluster, Cloud And Grid Computing (Ccgrid), 10th Ieee/Acm International Conference On. Ieee, 2010.

[14] Lin, Weiwei, James Z. Wang, Chen Liang, And Deyu "A Threshold-Based Dynamic Resource Allocation Scheme For Cloud Computing", Procedia Engineering, 2011.

[15] Yin, Bo, Et Al. "A Multi-Dimensional Resource Allocation Algorithm In Cloud Computing" Journal Of Information And Computational Science 2012.

[16] D. Pandit, S. Chattopadhyay, M. Chattopadhyay And N. Chaki,"Resource Allocation In Cloud Using Simulated Annealing," Applications And Innovations In Mobile Computing (Aimoc), Kolkata, 2014.

[17] Pandaba Pradhan , S. Pani , "A New Approach for Resource Allocation in Cloud Computing" , International Journal of Computer Science And Technology (IJCST), 2014.
[18] Li, Jiayin, Et Al. "Adaptive Resource Allocation For Preemptable Jobs In Cloud Systems." Intelligent Systems Design And Applications (Isda), 10th International Conference on. Ieee, 2010.
[19] Younge, Andrew J., Et Al. "Efficient Resource Management For Cloud Computing Environments." Green Computing Conference, Ieee 2010.
[20] Li, Jiayin, Et Al. "Online Optimization For Scheduling Preemptable Tasks On Iaas Cloud Systems." Journal Of Parallel And Distributed Computing 2012.
[21] Rammohan, N. R., And E. Baburaj. "Resource Allocation Using Interference Aware Technique In Cloud Computing Environment." International Journal Of Digital Content Technology And Its Applications 2014.
[22] Vishnupriya, S., P. Saranya, And P. Suganya. "Effective Management Of Resource Allocation And Provisioning Cost Using Virtualization In Cloud." Advanced Communication Control And Computing Technologies (Icaccct), 2014.
[23] Yuan, Haitao, Et Al. "Sla-Based Virtualized Resource Allocation For Multi-Tier Web Application In Cloud Simulation Environment." Industrial Engineering And Engineering Management (Ieem), 2012.
[24] Abdulkader, S. J., & Abualkishik, A. M. Cloud Computing And E-Commerce In Small And Medium Enterprises (SME's): The Benefits, Challenges.
[25] K Delhi Babu, D.Giridhar Kumar "Allocation Strategies Of Virtual Resources In Cloud Computing Networks" Journal Of Engineering Research And Applications.
[26] Son, Seokho, Gihun Jung, And Sung Chan Jun. "An Sla-Based Cloud Computing That Facilitates Resource Allocation In The Distributed Data Centers Of A Cloud Provider." The Journal Of Supercomputing 2013.
[27] Amazon, E. C. Amazon Elastic Compute Cloud (Amazon Ec2). Amazon Elastic Compute Cloud (Amazon Ec2), 2010.
[28] Minarolli, Dorian, And Bernd Freisleben. "Utility-Based Resource Allocation For Virtual Machines In Cloud Computing." Computers And Communications (Iscc), 2011.
[29] Buyya, Rajkumar, Anton Beloglazov, And Jemal Abawajy. "Energy-Efficient Management Of Data Center Resources For Cloud

Computing: A Vision, Architectural Elements, And Open Challenges." Arxiv Preprint Arxiv 2010.
[30] Ergu, Daji, Et Al. "The Analytic Hierarchy Process: Task Scheduling And Resource Allocation In Cloud Computing Environment." The Journal Of Supercomputing 2013.

7

Role of Machine Learning in Big Data

Nilanjana Das*, Santwana Sagnika and Bhabani Shankar Prasad Mishra

School of Computer Engineering, Kalinga Institute of Industrial Technology Deemed to be University, Bhubaneswar, Odisha – 751024, India
E-mail: {nilanjanadas010, santwana.sagnika, mishra.bsp}@gmail.com

Abstract

Machine learning is promising when large-scale data processing is considered as it has the capability of learning from the previous data and incorporating those findings on new incoming data. Although its applications in big data are limitless, it faces a vast number of challenges. This paper gives us an introduction to the implementation of machine learning in big data. We discuss about four out of the five Vs of big data and about the challenges faced in this regard. Some previous works are also discussed in the related work section. Some of the tools under batch analysis, stream analysis, and interactive analysis are explained here. Under the section of machine learning algorithms in big data, popular algorithms in supervised, semi-supervised, and unsupervised categories are briefly discussed. Machine learning also finds its way in sectors like healthcare, financial services, automotive, etc. It can also be used in chatbots, recommendation engines, user modeling, predictive analytics, etc. Then come the challenges that are faced by the machine learning algorithms. Under each characteristic, there come several areas that are a challenge and can be improved to enhance the productivity of the algorithm. Large and complex datasets are also difficult to process. New techniques need to be evolved or the existing technologies need to be updated for enhanced processing and better results.

Keywords: Big data, machine learning algorithms, tools and techniques, applications.

7.1 Introduction

With the onset of large volumes of data, traditional methods of data storage need to be enhanced. The authors in [1] define big data as a collection of very large complex datasets whose processing becomes difficult using the traditional applications for processing of data. This paper [1] also mentions that the biggest source of data, i.e., about 90% of the data, is from social media and also discusses the five Vs or the big data characteristics. The first one is the volume which refers to the complex and large quantity of data which is generated every second. The authors in [2] mentioned that in the near future, dataset will be generated in zettabytes which is far more than the capacity of processing of the current database systems. Next comes the velocity which is the speed of the generation of data. It is mentioned in [2] that it is the speed with which the data comes from the different sources and also the speed with which it flows. The authors in [3] referred velocity as the collection and analysis of data to be done in a timely manner. Variety of the data refers to the different categories of data available. It may be unstructured data or structured data. Value of the data refers to how much the data can provide us with relevant information. It is also mentioned in [3] that for the purpose of revenue, the data can be sold to third parties. Veracity of the data refers to both its validity and reliability. Veracity is referred in [3] as the quality of data which needs to be good, and proper sanitization and cleaning of data needs to be done so that the results we get are effective and accurate. Figure 7.1 represents the 5 Vs in big data. Coming to machine learning, it is a component of artificial intelligence and can make intelligent decisions. Machine learning algorithm can be of three types. They are supervised, semi-supervised and unsupervised. This paper mainly focuses on the introduction of machine learning in big data. With the increasing amounts of data, making the prediction models is becoming more difficult with the existing set of tools and techniques. Therefore, the authors in [4] mentioned the necessity to process large amounts of real-time data in a way that increases the accuracy of the predictions and can be incorporated into applications of different domains. Section 7.2 discusses on some of the previous works which implement machine learning in big data. Section 7.3 discusses some of the tools used in big data. Section 7.4 mentions some of the machine learning algorithms that are used in big data. Sections 7.5 and 7.6 list its applications and challenges, respectively. We finally end the paper with Sections 7.7 and 7.8 depicting the conclusion and references, respectively.

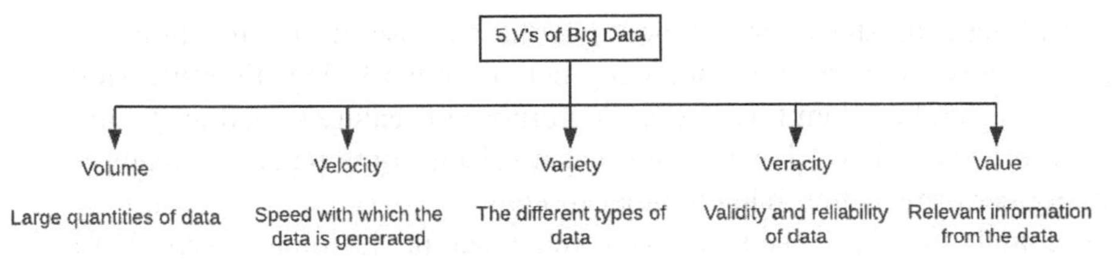

Figure 7.1 Five Vs of big data.

7.2 Related Work

Machine learning in big data provides very good solutions as it can perform well on large datasets. However there are several works in this field. The authors in [5] had made a detailed study on the applications, challenges and technologies in machine learning for big data. It also introduced some of the new methodologies for the same like multitask learning, incremental learning, faceted learning, and deep learning algorithms. The work [4] also explains some of the techniques like decision trees, artificial neural networks, cluster analysis, support vector machines, reinforcement learning, and genetic algorithms.

The authors in [6] used a deep reinforcement learning method which is a semi-supervised method for smart city implementation. Most of the real-time unlabeled data gets wasted. The paper implements this learning algorithm as it uses both the labeled dataset obtained from the feedback of the users and the unlabeled dataset which does not have the feedback of the users so that the wastage of unlabeled data can be minimized. The paper also brought into focus some of the possible challenges that may arise when machine learning is implemented in the services of smart city.

The paper [2] discussed the importance of big data in today's world and what the possible issues faced are while implementing them to projects. It also explained about the widely used big data tool Hadoop in detail. Another paper [7] discussed how machine learning combined with big data can be implemented in health systems, thus enhancing its discovery, performance, and prediction so that the organization's and patient's complexity can be reduced. Implementing machine learning with big data in health systems will require new or enhanced tools and techniques, training, and analytical thinking.

The work [8] discussed how machine learning with big data can be used in stock selection, i.e., to distinguish between good and bad stocks. To describe each stock, they make 244 fundamental and technical features. Based on their rankings, the stocks are labeled. For the purpose of classification, this paper [8] uses the algorithms like deep neural network, logistic regression, the stacking, and random forest. For the purpose of feature selection, genetic algorithm is used, but it is also seen that it selects some repetitive features. Stacking performs better than the other models.

The authors in [9] used big data and machine learning approach for estimating the travel time by collecting Taiwan electronic toll collection's real-time and historical data. Here, Apache Hadoop in combination with random forests was used to predict the travel time in the highway. This will guide the drivers with the estimated travel time that may be required by them. Two different models are used in the paper for different situations which took into consideration different sets of features having some features in common; some of them are traffic flow, gantry to, gantry from, vehicle type, mean speed, trip length, etc. The two models are adaptive travel time prognosis and one destination travel time prediction.

7.3 Tools in Big Data

Big data tools are categorized into categories based on their analysis as mentioned in [3]. They are batch analysis, stream processing, and interactive analysis. The data is first stored and then it undergoes analysis in batch analysis, whereas in stream processing, analysis is done as soon as possible so that the results can be derived quickly. Interactive analysis gives the users freedom to do their own information analysis. Apart from the above definitions, the paper also described a list of tools of big data under each category. They are as follows.

7.3.1 Batch Analysis Big Data Tools

- **MapReduce:** The paper [3] mentioned MapReduce as a JAVA-based framework which is a parallel computing model and is batch oriented. This paper [3] explained the strategy of MapReduce. It explained that there is a single master node which performs the function of how the Map workers will be assigned their work. Here the input data is divided into splits and then these splits are assigned by the master to the Map workers. Each Map worker processes its part of the assigned splits and then it generates the key/values pairs. Then it writes those key/value pairs to the intermediate files. Then the Reduce workers are informed by the master about the intermediate file's locations. The Reduce workers then read the key/value pairs from it, process it according to the reduce function and then write the final outcome to the output files. The paper also said that scalability, fault tolerance, and simplicity are the three main characteristics of MapReduce.

- **Apache Hadoop:** Let us discuss what [3] mentioned about Apache Hadoop. It is an open source software which is JAVA based. It has two components. One is Hadoop Distributed File System (HDFS) and the other is MapReduce. The process generally includes bringing the files into the distributed file systems and then executing the computations of MapReduce on the data. The work [3] also mentioned some of the programming languages used for computations in Hadoop which are R, Pig Latin, HiveQL, Python, and JAVA. The two sub-projects of Hadoop are HBase and Hive which give solutions for managing data to store unstructured and semi-structured data.
- **Microsoft Dryad:** It makes use of Dryad, a parallel runtime system and two other programming models as discussed in [3]. One is SCOPE language which is like SQL. Another is DryadLINQ. Let us discuss what else the paper had mentioned about Dryad. Dryad also gives the programmer the permission so that he can utilize the different computer cluster resources and the data parallel programs can be executed. Even without the knowledge of Dryad programming, a programmer of Dryad can utilize many machines. Channels which are one-way can be used to connect many sequential programs that are written by the programmer. The computations in Dryad are like a directed graph wherein the channels are the edges and the programs are like the vertices in a graph.
- **Apache Mahout:** It is an open source software which is used for machine learning and data mining tasks and is present in Hadoop [3]. Apart from this, the paper not only mentioned that it helps in building applications that are intelligent but also assists in three usages which are clustering, recommendation, and classification.

7.3.2 Stream Analysis Big Data Tools

- **Apache Storm:** It is a real-time open source and free computation system [3]. According to the paper, the processing speed of Apache Storm is very high, that is, a million records in a second per node can be processed and it can be used with any programming language. Thus, it is very simple and efficient to use.

- **Apache S4:** In Apache S4, computations are done in JAVA which is done by processing elements (PEs). It was also mentioned in [3] that the PEs can only interact with each other by using business events and one PE cannot know the status of the other PE. S4 can also properly route the events to the correct PEs. It is also fault tolerant, scalable, and a general purpose platform which enables the users to create application to process continuous streams of data.
- **Apache Spark:** It is an in-memory cluster computing framework [3]. It is open source and is a general purpose framework. It is used in large-scale industries. Some of its applications are traffic predictions and spam-filtering and has APIs in JAVA, Scala, and Python [3].
- **MOA:** It is also an open source software used for data mining purpose [3]. The paper also mentioned some of the areas of its applications which are clustering, classification, frequent graph mining, regression, and frequent set mining and another software project SAMOA which is a combination of S4 and Storm along with MOA used for mining of stream.

7.3.3 Interactive Analysis Big Data Tools

- **Apache Drill:** It can be used to process very large amounts of data, that is, data in petabytes and also large number of records per second whose numbers can be in trillions.
- **SpagoBI:** It is a business intelligence (BI) tool used for real-time analysis of big data. It gives meaningful insights and performs data management on unstructured and structured data by using tools like interactive dashboard, a business intelligence which is self-service, exploratory analysis, and *ad hoc* reporting [3]. Apart from this, the paper also discussed more about it. SpagoBI also uses a Query by Example Engine which enables the user to graphically create queries. It also has the functionality which enables the users to refine queries using the self-service BI. New datasets are created after many filter conditions are applied on it, and later on, it can be used for graphical analysis by *ad hoc* reporting engine.
- **D3:** The data present in different types of documents can be manipulated using D3 which is a library based on JavaScript [3]. The paper also discussed that it helps in visualizations of different types of datasets and it makes use of HTML and CSS rather than the expensive design tool like Adobe Photoshop, thus making it browser friendly and easy to

change the styles using CSS and also making it easily available to users through the browser. It also helps in picking up data visualizations that are suitable.

7.4 Machine Learning Algorithms in Big Data

Let us discuss what [10] had discussed about the different machine learning algorithms in big data in his paper. There are two types of machine learning algorithms. They are discriminative and generative algorithms. In a discriminative algorithm when a model is to be learnt, prior distribution of data is not taken into consideration, whereas generative algorithms depend on distribution of data for model learning. These algorithms have been explained with an example. Suppose we have two classes of fruits like oranges and apples where y is the predicted output 0 and 1, respectively, and x is the set of features for the fruits. In this case, discriminative method will try to find out a decision boundary which will clearly separate both types of fruits and it knows which side is for apples and which side of the decision boundary is for oranges and if any new fruit, which is to be classified, is first seen on which side it is present, and accordingly, a prediction for that fruit is made. But a generative method makes two models, one for apple and one for oranges. Then a new fruit is checked against those two models and then labeled as that fruit with whose model it resembles the most. Both [10] and [11] mentioned that the discriminative model has lesser asymptotic error than the generative model. Some of the generative models are Naïve Bayes, Markov random fields, hidden Markov models, Bayesian networks, etc. Some of the discriminative models are neural networks, support vector machines (SVMs), logistic regression, etc. These two algorithms together perform a particular task. Table 7.1 discusses the various machine learning algorithm types.

- **Supervised learning for big data:** There are also a number of supervised machine learning algorithms discussed in the paper [10]. Now let us discuss what the paper had said about this. In supervised machine learning, an input labeled dataset is used to train an algorithm so that it can predict the labels of the new incoming data. The output y is a function of x, i.e., $y = f(x)$, where x is the input training dataset. There are two types of supervised machine learning algorithms. They are regression and classification methods. Classification methods can be used to classify x into classes that are discrete, whereas regression

Table 7.1 Overview of different types of algorithms.

Supervised learning	Labeled dataset is required.
	Follows two types of approaches for learning: classification and regression.
	Algorithms are logistic regression, linear regression, SVM, neural networks, etc.
	Costly as large number of human annotators are required to label the dataset.
Unsupervised learning	Unlabeled dataset is required.
	Follows two types of approaches for learning: clustering and association.
	Algorithms are principal component analysis, K-NN, etc.
	Does not require human annotators as unlabeled dataset is used and it is cost effective.
Semi-supervised learning	Partially labeled and unlabeled dataset is required.
	Algorithms are label propagation, co-training, etc.
	Requires very less number of human annotators as most of the data available is unlabeled and is suitable for the problems of the real world.

method is used to predict dependent variables after the independent variables are modeled. Let us discuss some of the supervised machine learning algorithms.

- **Decision tree:** This has a structure of a tree. It basically starts its evaluation from the roots and continues to the leaf nodes and gets its final output and hence classifies the particular sample. There are three variations of decision trees that are being used today. They are C4.5, ID3, and CART. Figure 7.2 gives an example of a simple decision tree which says that a person is only eligible for discount if he is a member of that group and if his age is greater than 60.

- **Logistic regression:** This algorithm implements a model that can be used to find a relation between independent variables and dependent variables. The independent variable can be ordinal, nominal, etc., whereas the dependent variables are simply binary numbers 0 or 1. The paper also plotted a graph and gave Equation (7.1) for logistic regression problems. The equation is as follows:

$$y = \frac{1}{1 + e^{-(\alpha+\beta x)}}. \tag{7.1}$$

- **Regression and forecasting:** The regression analysis is used to calculate the statistical relationships present between the variables and

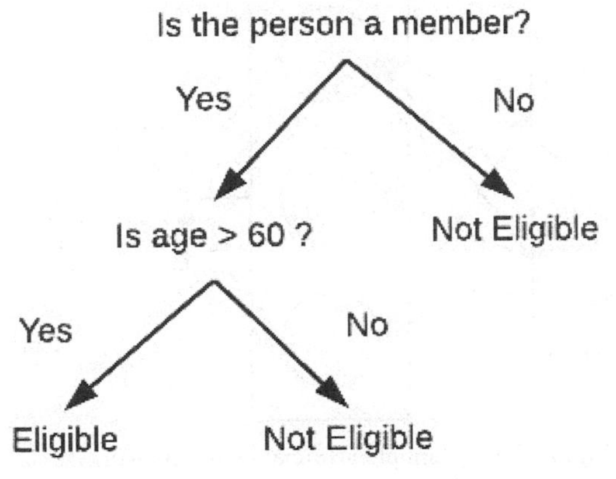

Figure 7.2 Eligibility of a person for discount.

getting the model parameters so that it can predict the future data values.

- **Neural networks:** It is basically the concept of human brain that can be artificially implemented. They are useful in large-scale processing and can be enabled on architectures of multiple processors. It can be used to predict the final outcome based on the training it gets and once it is able to recognize the input pattern. Similar to the human brain, its unit of computation is an artificial neuron which, with a weight of w, gets the other units' input. It also gave Equation (7.2) to calculate the function f of the inputs' weighted sum which is written below. f is an activation function which limits the output's amplitude to a particular fixed range and can also be used to determine a neuron's characteristics. Many neurons together form multiple or a single layer. The main duty of the neural network is to learn the weights from inputs and each time calculation of an error is done so that the weights can be changed accordingly and can fit the model accurately. Figure 7.3 represents the basic architecture of a simple artificial neural network.

$$f\left(\sum_{j} w_{ij} y_j\right) \tag{7.2}$$

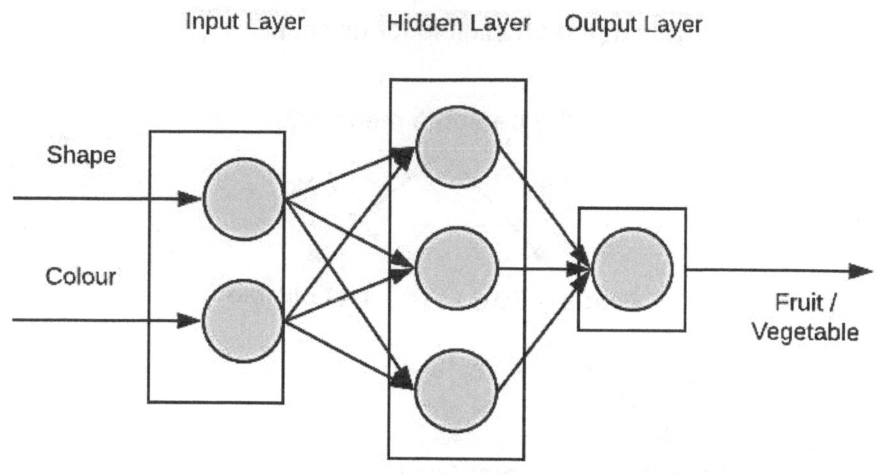

Figure 7.3 A simple artificial neural network.

- **Support vector machines:** This is a discriminative algorithm where a decision boundary, i.e., an optimal hyperplane is used to separate two classes.
- **Unsupervised learning for big data:** Unsupervised machine learning algorithms are not trained on a labeled dataset. We only have x here and no y is present. This includes two types of methods, association and clustering. Clustering groups similar kind of input data together into a single cluster. The association method is used to identify rules that fit to a large volume of input data. Since it does not need the output labels, it passes the overhead of annotating large amounts of data and is suitable for big data.
- **Spectral clustering:** K-NN and K-means are some of the available clustering methods. However, spectral clustering is one of the most promising methods of clustering. It is based on graphical Laplacian matrices, where G is the weighted graph, W is the weight matrix, and $w_{ji} = w_{ij} \geq 0$. The matrix eigenvectors are not normalized. There are three steps in spectral clustering. For the N number of objects which are to be grouped together, a similarity graph is to be made. To get an object's feature vector, the Laplacian matrix's first k eigenvectors are calculated. We finally get the k classes of the objects by running the k-means algorithm on the features.
- **Principal component analysis (PCA):** This uses matrix factorization and is used to reduce the number of dimensions by considering only

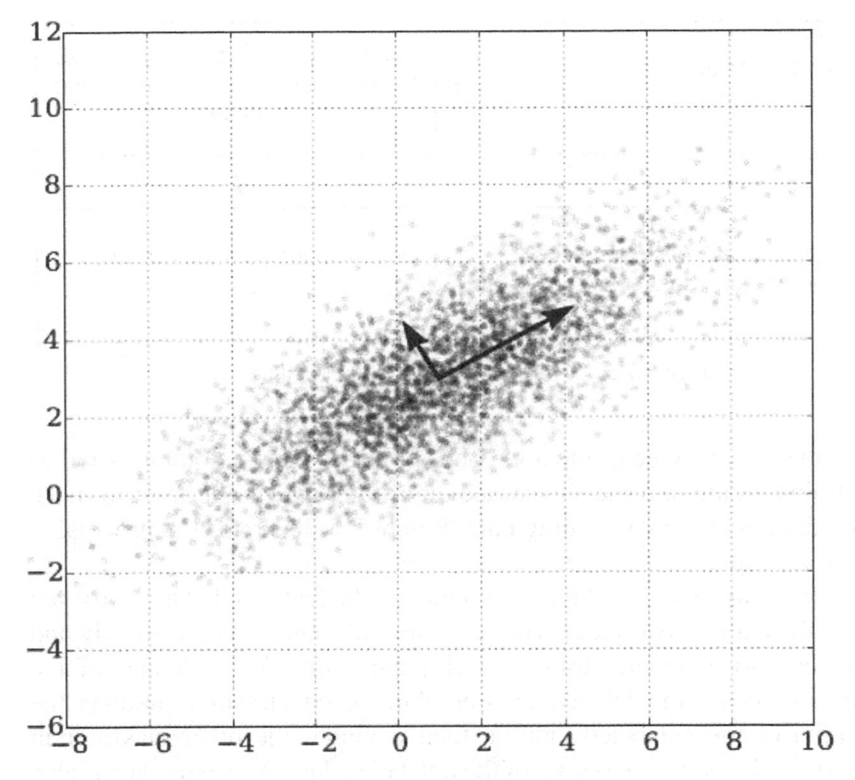

Figure 7.4 A multivariate Gaussian distribution PCA [12].

the relevant components which are present in the vector space and also converts correlated variables into many smaller principal components i.e., uncorrelated variables which are responsible for data variability. PCA applies the concept of factor model which takes into account a shared variance. Thus, the correlation matrix's diagonal has unities. Although it is a linear model, it can also be used for non-linear modeling. Figure 7.4 illustrates the example of a multivariate Gaussian distribution PCA [12].

- **Semi-supervised learning for big data:** Sometimes, it may happen that a very large input dataset is not fully labeled. It applies both the methods of learning of supervised and unsupervised processes. This is very efficient and can fit the real-world problems today as most of the data in the real-world applications is unlabeled and it may cost very high to label the data using human annotators. It is easy to get

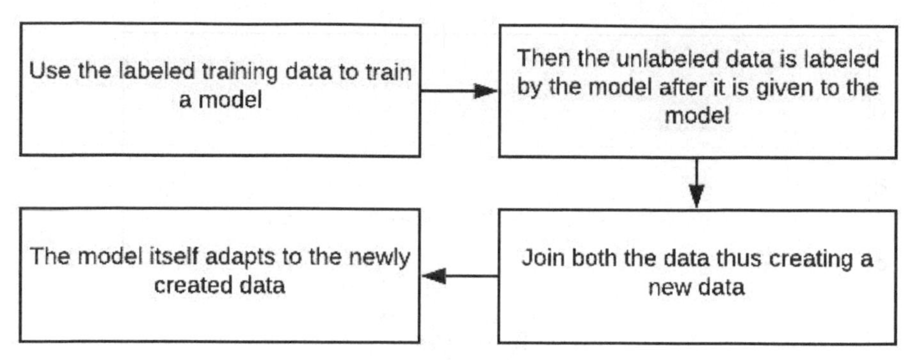

Figure 7.5 Semi-supervised machine learning.

unlabeled data. One of the approaches it uses is that its output is fed as a training data into the supervised algorithm and can be used to make predictions on new incoming data. Figure 7.5 illustrates the processing of semi-supervised learning.

- **Co-training:** This has three assumptions. The features that are extracted are divided into two parts. The parts are independent conditionally and are able to efficiently train two classifiers, one for each part of the features using the labeled dataset. After each classifier predicts the output of the unlabeled dataset, then it guides the other classifier in learning the predicted output of the unlabeled data. Any extra data is also used to train the classifiers and this cycle continues unless the required accuracy is reached and there is a proper agreement among the classifiers on the labeled and the unlabeled data. Figure 7.6 has been illustrated in [13], which shows the process of co-training.
- **Label propagation:** It is an algorithm which is iterative in nature. This follows the concept that when there are more unlabeled data, the labels are then propagated there. K-NN can be used for this purpose, but better algorithms than this are required to be found out for propagating the labels. This paper [10] also mentioned two methods, parameter optimization based on entropy minimization and heuristic of minimum spanning tree used to determine the proximity of the nodes on the basis of which the labels are propagated. But the algorithm does not have the power over class proportions; hence, this does not promise an equal class distribution. Including constraints in the proportions of the classes ameliorates the classification when there is less labeled data present. Figure 7.7 has been shown in [14].

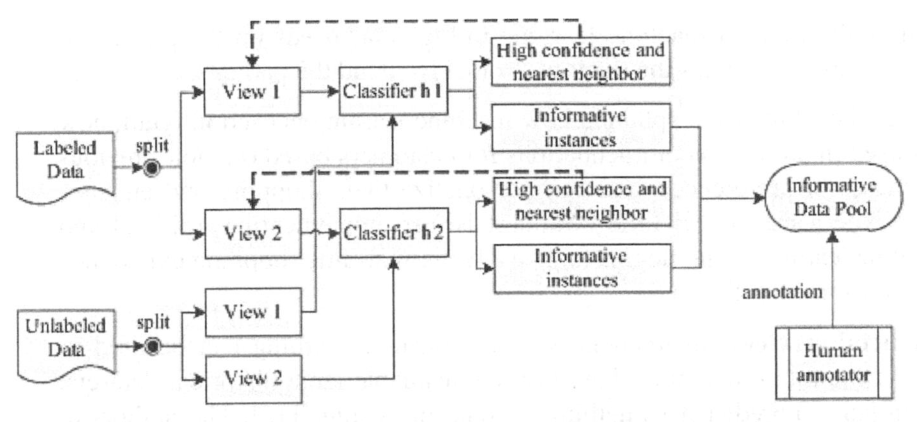

Figure 7.6 Co-training algorithm [13].

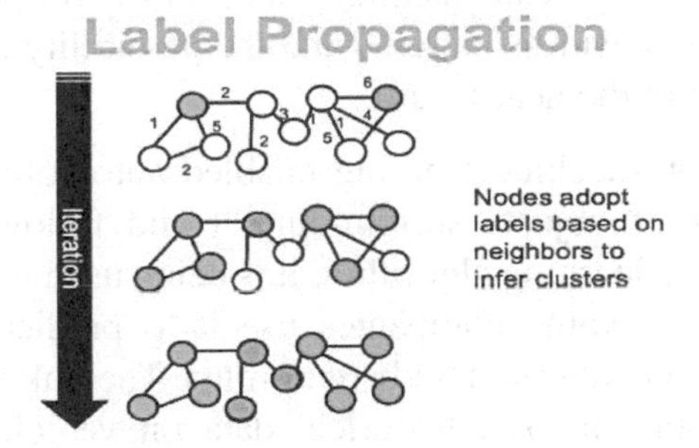

Figure 7.7 Label propagation [14].

7.5 Applications of Machine Learning in Big Data

With the vast amount of data being produced per second, analyzing and managing them has become a problem these days as mentioned in the link [15]. This link also says that when big data is combined with machine learning, it gives better results in comparison to when only big data is used. This increases not only the productivity of the companies, but also they can now concentrate more on catering the needs of their customers. The paper [15] listed a number of applications of machine learning in big data. They are as follows.

Healthcare: Machine learning has been used in treatment to predict many diseases such as cancer. Machine learning has been able to analyze vast amount of medical data to predict if the person has a disease or not and also analyze the pattern of the data [15]. The above link also mentioned that such applications of machine learning in big data speeds up the process of diagnostics and enables the doctors to quickly attend the patients.

Retail: This link also explained how machine learning is used in retail, how it is used in product recommendations for customers based on their previous purchases or preferences, and can personalize their shopping and enhance their experience in real time. Walmart is also implementing artificial and machine learning technologies to give a better real-time shopping experience to its customers [15].

Financial services: In financial sectors, machine learning can be used to predict the future outcomes based on the available large historical datasets. It can help to predict if something is fraud or genuine [15]. The authors in [15] mentioned that it helps to prepare strategies so that financial sectors can invest in those areas where it can maximize its profit and opportunities. It also helps to analyze the business patterns and the possibility of a particular pattern that can be used in the near future.

Automotive: Nowadays, machine learning enabled automobiles are tested which analyze the list of objects surrounding it and follow the path as suggested by the machine learning algorithm. It is being used in many sectors of automobile industries. Some companies use it to predict as to which marketing strategies can be inherited for better profits. The link also explained that by implementing this in past historical data on vehicle ownerships, business with which dealer can be initiated for increased profits. It can also be used to predict if the vehicle has any parts that may fail in the near future so that there is less need of expending on maintenance.

According to [16], machine learning labels a dataset and identifies the patterns in it which can then be used to improve business strategies. They are generally used to automate different decision-making processes. Machine learning becomes very important to be used by organizations that generate very large amount of data. In addition to the above, [16] also lists three areas of big data where machine learning is used in an organization. They are:

- Data labeling and segmentation
- Data analytics
- Scenario simulation

The work [16] also listed some of the machine learning in big data projects. Let us discuss how [16] describes them one by one.

Market research and target audience segmentation: Many supervised and unsupervised machine learning algorithms are used for this case. Like in [15] and [16], it is also explained that such algorithms can be used to analyze the behavior of the audience and predict what related products they might like on the basis of their preferences and thus serving them in an improved manner.

User modeling: This follows the target audience segmentation. Similar to the previous one this is also used to predict and analyze the behavior of the users to improve the business strategies. Such an example is Facebook which has a complicated user modeling system, which is used to predict friends, pages, ads, etc. to users after analyzing their profiles [16].

Recommendation engines: The paper [16] explained that it follows the principle of content-based data extraction from different e-commerce sites. It is similar to the previous usages. It suggests the users, products on the basis of their preferences, and interests.

Predictive analytics: The paper [16] used a figure to explain predictive analytics. The figure demonstrates that a model is trained using predictive algorithms on historical data. Then the trained model is used to make predictions on new data. According to [16], predictive analytics is used in three different areas. First, it is used in e-commerce sites to suggest some extra products. Second, it is used to see if any fraudulent activity is possible or not in an ad tech project. Third, it can be used in healthcare to see if there is any probability of increasing efficiency of treatment of patients. It also gives an example of eBay which uses predictive analytics for its business operations.

Ad fraud and e-commerce fraud: According to [16], 10%–30% advertising activities are fraud. It also explicitly mentioned three uses in this category. First, it can identify the patterns in big data. Second, their credibility can be assessed. Third, blocking the fraudulent activities from the system before anyone tries to access it, be it a bot or any user.

Chatbots: Chatbots can nowadays answer many personalized questions from different users based on their choice of topic of interest. Some of the most famous personal assistants today are Siri by Apple and Alexa by Amazon. Machine learning in big data will improve the workforce. Machines do not have human intelligence, so they will work alongside humans to improve the efficiency of the programs [17]. According to [17], it will help the companies to globally diversify and find better business solutions.

7.6 Challenges of Machine Learning in Big Data

This section discusses the various challenges faced while applying machine learning to big data. The papers [18] and [19] discussed the characteristics of big data and the machine learning problems faced by them. Let us discuss what the papers mentioned about them one by one. Big data has five characteristics, out of which they discussed four characteristics and their challenges. The four characteristics discussed by them are volume, variety, velocity, and veracity. These characteristics have already been discussed earlier. There are a number of areas under each characteristic which face machine learning challenges. Let us discuss what they had explained about the challenges in each characteristic at a time.

7.6.1 Volume

The following points have been discussed by [18] under volume:

- The performance of processing can be affected as processing large volumes of data using machine learning can lead to an increase in time and space complexity.
- To process data iteratively or parallelly is a machine learning challenge.
- Data can become corrupted if different techniques of sampling are applied to it.
- Reliability will decrease if the number of dimensions increases. The same goes for effectiveness also.
- When data volume increases, the cost also increases as the appropriate features are extracted.
- Another challenge for machine learning is also identifying non-linearity.
- When the volume of data increases, overfitting problems may arise.

All the above have been discussed in [18]. The paper [19] also discusses the same challenges as in [18]. The following points have been discussed by [19] under volume:

- Large volumes of data increase the computational complexity. Because of the large volumes of data, even computations of lesser significance can become costly. The paper [19] also cited an example with this regard. The time complexity for training of SVM model is O(m^3) and the space complexity is O(m^2). The number of training examples is m. Now if m becomes very large, then it is difficult to train the SVM algorithm as it will take up much space and time. Not only SVM, this paper also mentions other algorithms with their time complexities. O $\left(mn^2 + n^3\right)$

is the time complexity of PCA, linear regression, logistic regression, and Gaussian discriminative analysis. The number of features is n. Here, if the data size increases, then the computation time will also increase exponentially. In such cases, the algorithms will become futile. Partitioning and placement of data, parallel data structures, and the reuse of data become important factors to be considered with the increase in size of data. Resilient Distributed Datasets (RDDs) RDDs are used for in - memory computations.

- The need for parallel execution can be solved by using MapReduce which can simultaneously execute many numbers of nodes. The work [19] had also mentioned three methods used to counter the challenge of course of modularity. They are gradient descent, iterative graph, and expectation maximization algorithms. Since these algorithms depend on the in-memory data and also they are iterative in nature, together they create a disconnect with MapReduce's distributed and parallel nature. This makes it difficult to adapt such algorithms to MapReduce. However k-means algorithm can be used to overcome this challenge.
- There may also arise a problem of class imbalance where many samples can represent a single class, whereas very few samples may represent a single class. Training of algorithm with such type of large dataset becomes difficult and gives inaccurate results. Both [19] and [20] mentioned three algorithms which are class imbalance sensitive; they are neural networks, decision trees, and SVM.
- Number of dimensions is the number of features. With the increase in the number of features, the accuracy and performance of machine learning algorithms decrease.
- Then there comes feature engineering where relevant features are generated which will help reduce the dimensionality. Feature selection also helps in selecting the appropriate features from the dataset and thus reduces the time complexity of machine learning algorithms.
- The paper explains the challenges further with an example of non-linearity. There exists a correlation coefficient which is a sign of good indicator between two or more variables, but this is only accountable if there exists a linear relationship between the two variables; otherwise, it is futile. Deciding whether the plot is linear or non-linear using scatterplots becomes difficult when the data size is very large, and hence, there are too many points in the plot.
- Another challenge is based on Bonferonni's principle which says that the probability of finding a particular event within a specific amount

of data is high. Most of the time, these occurrences are futile and of no significance. This principle helps in reducing the number of such futile searches. In addition to this paper [19], [21] has also mentioned a significance level (α/m) to be calculated, where m is the number of hypothesis to be tested with a significance factor of α. However, the chances of getting these events become higher out of which many are spurious, also mentioned by [22]. Removing these false positives becomes necessary to be discussed in the area of machine learning algorithms with big data.

- Machine learning follows the generalization idea. It consists of two parts, the bias and the variance. The consistency of the ability of a learner for the prediction of random things is called variance. Bias is the learner's ability to learn the wrong things. Values of both should be minimum so that we get a correct output. When the data volume increases, the tendency of becoming bias to the training set also increases and may not be able to fit well to the new data. Regularization technique can be used to reduce the problem of overfitting and enhance regularization. However, how much weightage regularization carries with respect to big data needs to be analyzed.

7.6.2 Variety

The following points have been discussed by [18] under variety:

- Storing large volumes and varieties of data, structured or unstructured in local storage, is difficult to provide and cannot be guaranteed.
- Uniformly distributing the data across the many available data sources is difficult.
- Costs increase when the missing data is to be handled and the data cleaning is to be performed.

The following points have been discussed by [19] under variety:

- Variety of a dataset not only takes into consideration the data types it contains in the dataset but also the dataset's structural variation. Both [19] and [23] said that machine learning algorithms assume that data is stored in a single disk file or in memory. In case of big data, the data is not stored in a single place; rather the data is distributed into different files among different locations. Then all the data has to be brought to a single location which will be very hectic if the data size is very large and will lead to very high network traffic. Thus, to solve this, a new approach that is bringing computation to model is followed rather than bringing data to computation. This is cheaper. MapReduce follows the approach of bringing computation to data. The map tasks perform the functions on the nodes where the data is residing.

- Data heterogeneity is another challenge faced. The two main categories of big data heterogeneity are semantic and syntactic heterogeneity. Syntactic heterogeneity is the variety of different formats of files, different types of data, data model, and encoding. Every data undergoes a pre-processing and cleaning method so that it is able to fit a model. These data come from various data sources; thus, they are of different types. Semantic data is the different types of interpretations and meanings of data. Machine learning was not made by keeping in mind the challenge of data heterogeneity. Hence, a solution to this challenge needs to be found out.
- Now coming to the challenge of dirty and noisy data, both [19] and [24] mentioned that data has three features which can be used for the purpose of characterization. They are condition, location, and population of the data. It takes a lot of time to make data analysis ready. Sometimes big data is also thought of as dirty as it comes from different locations and it also may have populations that may be unknown. It is also a challenge to treat outliers and the measurement errors as mentioned by both [19] and [25]. The noise also increases when the dimensions increase with the increase in data. Both [19] and [26] mentioned that after the data is collected and integrated, there should be a method of extracting signals from the noise. They also mentioned that big data cannot produce results that are meaningful as it is too noisy.

7.6.3 Velocity

The following points have been discussed by [18] under velocity:

- To a stream data, one cannot apply the train, test, and predict method and changes always find their ways in the model.
- In real-time data streaming, it is a challenge to reduce the time required to train the models.
- Here the models are always subject to changes, what may be good now may become worse later.
- Overall population cannot be assured by statistically independent variables.

The following points have been discussed by [19] under velocity:

- The velocity of big data is not only the speed with which the data is being generated but also the speed with which the data is being analyzed. A big data challenge is data availability where the data is always assumed to be present, but with the real-time scenario, data can come at any time and may come even at non-real-time scenarios. In view of machine learning, a model needs to be trained using the existing data present and then make predictions on new incoming data, but to make accurate predictions, a model needs to be retrained using the new data also. This again becomes a challenging situation as there is no limit to new incoming data. To counter this, both [19] and [27] suggested to use incremental learning also known as sequential learning that does not require the model be retrained on the entire data, instead keep on learning with the new incoming data. This algorithm works by keeping in mind that the entire data is not present and new data may add up in later stages.
- Real-time processing is much like the challenge of data availability. It is a challenge to constantly process real-time data. The subtle difference between the two is that in data availability, there is a necessity to train the model on the basis of the incoming data, whereas real-time processing focuses on the necessity to handle real-time data. Some real-time systems as mentioned by [19], [28], and [29] are Twitter's Storm and Yahoo's S4. These two systems do not use very complicated machine learning algorithms but still have succeeded a lot in real-time processing of data. If machine learning could be integrated with these systems, then the results will be better. Thus, it is impossible to process and analyze the entire dataset as we do not have it at any given point of time.
- Concept drift is another challenge as mentioned in the paper. It mainly refers to the fact that the conditional target distribution may change, given the input distribution remains constant. This is a problem as the machine learning model is already fitted on older set of data and is not accurately fitted on newer set of data. This issue can be solved using sliding window concept. This concept uses the training data to train the model, where the training data includes the new incoming data. In this way, the model is constantly trained on the latest incoming data. Assuming that the latest data is always important is not true always as mentioned in both [19] and [30]. Both [19] and [31] mentioned the different concept drift types which are sudden, gradual, incremental, and

recurring. The onset of big data made many existing methods futile due to the rise in the presence of the issue of concept drift. Thus, for a better future of machine learning, this issue of concept drift with respect to big data needs to be handled.

- Another issue is the independent and identically distributed random variables. It basically assumes the fact that all the random variables are mutually independent of each other and that all of them have the same probability distribution. But this may not be true in reality. In addition to this, different algorithms depend on different distributions. Big data has a set of non-random data which contrasts the challenge of independent and identically distributed random data. An attempt can be made to randomize the data before applying machine learning algorithms, but taking into account big data, this itself becomes a challenge. However, it is not recommendable to randomize data without having the entire data. Having the entire data at once is also not possible as there is a constant inflow of data.

7.6.4 Veracity

The following points have been discussed by [18] under veracity:

- Metadata collected and administered for configuring different machine learning techniques thus increases the costs.
- It is uncertain whether the sentiment analysis performed on social media is giving correct results or not.
- Costs also increase when the data quality is poor.

The following points have been discussed by [19] under veracity:

- Both [19] and [32] described data provenance as recording and keeping a track of where the data originates from and where it propagates to from the locations. The resultant metadata that is captured is the contextual information that can be useful to machine learning. However, in the view of machine learning, this metadata is itself very large and becomes a challenge in itself as mentioned by both [19] and [33]. A MapReduce method namely, Reduce and Map Provenance, has been developed as a Hadoop extension for data provenance [34], but this only increases the cost and complexity of implemented machine learning in big data [19].
- The veracity depends a lot on the certainty of data. But the different ways of data collection may not always guarantee data and their veracity cannot be justified with an incomplete set of data. Data from social

media and crowdsourcing are the two new methods of gathering large amounts of data. Crowdsourcing as the name suggests can be used to obtain ideas from a large crowd of people. This data gathered has a certain degree of uncertainty. This type of imperfect data cannot be handled by machine learning, thus, again leading to the challenges of machine learning with big data.

- Another challenge is dirty and noisy data which becomes difficult to remove when it comes to large volumes of data. The contextual information or the data labels may not be correct. Such a condition is different from imperfect data from the point of view of machine learning. Dirty and noisy data may come from a variety of sources. Uncertainty can be caused by crowdsourcing when it is applied in participatory sensing but can also cause noisy data when man labels the dataset. Inaccurate assignments can be made either by mistake or on purpose. This wrong labeling impacts the veracity of data and can also hamper the training of machine learning models with wrong data labels. New methods need to be thought of as how this issue can be tackled when machine learning is applied to big datasets.

The paper [6] brings out an approach to implement smart city applications. However, it also listed and explained some of the challenges while implementing machine learning in big data. New real-time analytic frameworks will be required by joining both streaming analytics and big data analytics as large inflow of real-time data will occur that have to be acted upon in a limited amount of time. As the machine learning algorithms will be working based on incoming data, it so happens that false data is injected, and thus, the trustworthiness of the results become questionable. Also most of the data is of private individuals who would like to keep their data secured, thus protecting the privacy of these individuals also becomes a challenge. Availability of real-time datasets is a problem too. There is a necessity of properly integrating both raw data and contextual information to provide correct reasoning. There is also a scarcity of comprehensive testbeds.

7.7 Conclusion

This paper is an introduction of machine learning to big data. It discusses a lot of big data tools that are today used by many IT industries. Then it discusses some of the machine learning algorithms and their implementation in big data where we can see that neural networks work well with big data as the neural network is trained well because it has so many layers. Machine learning is also used in many sectors like retail, healthcare, financial services, etc., which is mentioned in the applications of machine learning in big data. The challenges discuss the different criteria's under each of the four Vs – volume, variety, velocity, and veracity – which have been discussed in the paper and pose a challenge to machine learning in view of big data. Discussing the challenges brings into focus the areas which can be improved for a better performance of machine learning. This paper will help other researchers to get an overview of machine learning in the context of big data and develop future research works. This will help the researchers to compare the algorithms and implement one of those which might give better results and also avoid one of the possible issues they may face while implementing it.

References

[1] Iza Moise, Olivia Woolley Meza, Lloyd Sanders, Nino Antulov- Fantulin (2017) Big Data and Machine Learning. https://ethz.ch.

[2] Katal, A., Wazid, M., & Goudar, R. H. (2013, August). Big data: issues, challenges, tools and good practices. In 2013 Sixth international conference on contemporary computing (IC3) (pp. 404-409). IEEE.

[3] Rodríguez-Mazahua, L., Rodríguez-Enríquez, C. A., Sánchez-Cervantes, J. L., Cervantes, J., García-Alcaraz, J. L., & Alor-Hernández, G. (2016). A general perspective of Big Data: applications, tools, challenges and trends. The Journal of Supercomputing, 72(8), 3073-3113.

[4] Nita, S. L., Dumitru, L., & Beteringhe, A. (2015). MACHINE LEARNING TECHNIQUES USED IN BIG DATA.

[5] Wang, L., & Alexander, C. A. (2016). Machine learning in big data. International Journal of Mathematical, Engineering and Management Sciences, 1(2), 52-61.

[6] Mohammadi, M., & Al-Fuqaha, A. (2018). Enabling cognitive smart cities using big data and machine learning: Approaches and challenges. IEEE Communications Magazine, 56(2), 94-101.

[7] Krumholz, H. M. (2014). Big data and new knowledge in medicine: the thinking, training, and tools needed for a learning health system. Health Affairs, 33(7), 1163-1170.

[8] Fu, X., Du, J., Guo, Y., Liu, M., Dong, T., & Duan, X. (2018). A Machine Learning Framework for Stock Selection. arXiv preprint arXiv:1806.01743.

[9] Fan, S. K. S., Su, C. J., Nien, H. T., Tsai, P. F., & Cheng, C. Y. (2018). Using machine learning and big data approaches to predict travel time based on historical and real-time data from Taiwan electronic toll collection. Soft Computing, 22(17), 5707-5718.

[10] Prabhu, C. S. R., Chivukula, A. S., Mogadala, A., Ghosh, R., & Livingston, L. J. (2019). Big Data Analytics. In Big Data Analytics: Systems, Algorithms, Applications (pp. 195-215). Springer, Singapore.

[11] Ng, A. Y., & Jordan, M. I. (2002). On discriminative vs. generative classifiers: A comparison of logistic regression and naive bayes. In Advances in neural information processing systems (pp. 841-848).

[12] https://en.wikipedia.org/wiki/Principal_component_analysis.

[13] Zhang, Y., Wen, J., Wang, X., & Jiang, Z. (2014). Semi-supervised learning combining co-training with active learning. Expert Systems with Applications, 41(5), 2372-2378.

[14] https://twitter.com/amyhodler/status/1114332199778758656.

[15] Kerri Hale (2018). Machine Learning and Big Data — Real-World Applications. https://towardsdatascience.com/machine-learning-an d-big-data-real-world-applications-3ba3a3345cf5.

[16] https://theappsolutions.com/blog/development/machine-learning-and-big-data/

[17] Gunjan Dogra (2018). Big Data Analytics Using Machine Learning Algorithms. https://nationalinterest.in/big-data-analytics-using-m achine-learning-algorithms-c33ef8488638.

[18] Augenstein, C., Spangenberg, N., & Franczyk, B. (2017, November). Applying machine learning to big data streams: An overview of challenges. In 2017 IEEE 4th International Conference on Soft Computing & Machine Intelligence (ISCMI) (pp. 25-29). IEEE.

[19] L'heureux, A., Grolinger, K., Elyamany, H. F., & Capretz, M. A. (2017). Machine learning with big data: Challenges and approaches. IEEE Access, 5, 7776-7797.

[20] Japkowicz, N., & Stephen, S. (2002). The class imbalance problem: A systematic study. Intelligent data analysis, 6(5), 429-449.

[21] Dunn, O. J. (1961). Multiple comparisons among means. Journal of the American statistical association, 56(293), 52-64.

[22] Calude, C. S., & Longo, G. (2017). The deluge of spurious correlations in big data. Foundations of science, 22(3), 595-612.

[23] Parker, C. (2012, August). Unexpected challenges in large scale machine learning. In Proceedings of the 1st International Workshop on Big

Data, Streams and Heterogeneous Source Mining: Algorithms, Systems, Programming Models and Applications (pp. 1-6).

[24] B. Ratner (2011) Statistical and Machine-Learning Data Mining: Techniques for Better Predictive Modeling and Analysis of Big Data. CRC Press.

[25] Fan, J., Han, F., & Liu, H. (2014). Challenges of big data analysis. National science review, 1(2), 293-314.

[26] Swan, M. (2013). The quantified self: Fundamental disruption in big data science and biological discovery. Big data, 1(2), 85-99.

[27] X. Geng and K. Smith-Miles (2019) "Incremental Learning," in Encyclopedia of Biometrics SE - 304, Springer US, pp. 731–735.

[28] N. Marz (2014) "Apache Storm,". [Online]. Available: http://storm.apac he.org.

[29] Neumeyer, L., Robbins, B., Nair, A., & Kesari, A. (2010, December). S4: Distributed stream computing platform. In 2010 IEEE International Conference on Data Mining Workshops (pp. 170-177). IEEE.

[30] Gama, J., Žliobaitė, I., Bifet, A., Pechenizkiy, M., & Bouchachia, A. (2014). A survey on concept drift adaptation. ACM computing surveys (CSUR), 46(4), 1-37.

[31] Dongre, P. B., & Malik, L. G. (2014, February). A review on real time data stream classification and adapting to various concept drift scenarios. In 2014 IEEE International Advance Computing Conference (IACC) (pp. 533-537). IEEE.

[32] Buneman, P., Khanna, S., & Tan, W. C. (2000, December). Data provenance: Some basic issues. In International Conference on Foundations of Software Technology and Theoretical Computer Science (pp. 87-93). Springer, Berlin, Heidelberg.

[33] Wang, J., Crawl, D., Purawat, S., Nguyen, M., & Altintas, I. (2015, October). Big data provenance: Challenges, state of the art and opportunities. In 2015 IEEE International Conference on Big Data (Big Data) (pp. 2509-2516). IEEE.

[34] Park, H., Ikeda, R., & Widom, J. (2011). Ramp: A system for capturing and tracing provenance in mapreduce workflows.

[35] Sagnika S., Mishra B.S.P., Dehuri S. (2016) Parallel GA in Big Data Analysis. In: Mishra B., Dehuri S., Kim E., Wang GN. (eds) Techniques and Environments for Big Data Analysis. Studies in Big Data, vol 17. Springer, Cham. https://doi.org/10.1007/978-3-319-27520-8_6.

8

Healthcare System for COVID-19: Challenges and Developments

Arun Kumar[1] **and Sharad Sharma**[2]

[1]Panipat Institute of Engineering and Technology, Samalkha
[2]*Maharishi Markandeshwar* (Deemed to be University), *Mullana*
E-mail: ranaarun1.ece@piet.co.in, sharadpr123@rediffmail.com

Abstract

The Internet of Healthcare Things is, in a general sense, changing how individuals speak with one another and with their general surroundings. Internet of Things (IoT) is the propelled organized framework of the network, transportation, and innovation. IOT brilliant gadgets can execute the offices of remote wellbeing checking and crisis notice framework. IoT has obvious utilization of shrewd medicinal services framework. In the social insurance framework, the featured approaches and methodologies help the analysts, researchers, and specialists who create keen gadget which is the up-degree to the current innovation. Associated wellbeing is a model for human services conveyance that uses computerized interchanges and the Internet of Healthcare Things to give quality medicinal services outside the customary emergency clinic setting. Associated wellbeing gives adaptable chances to patients to draw in with social insurance suppliers, frequently utilizing promptly accessible shopper advances. Be that as it may, conventional associated wellbeing systems are predominately centered on the customer end of the market, instead of adopting an incorporated and consistent strategy to transitional medicinal services from home, to outpatient, and to in-patient settings. This chapter provides some solution developments with challenges of IoT in healthcare for COVID-19.

Keywords: Internet of Things (IoT), Internet of Healthcare Things (IoHT), COVID-19, challenges and developments.

8.1 Introduction

China has reacted swiftly by informing the World Health Organization (WHO), after the discovery of the potential cause, of the outbreak and sharing sequence data with the international community. The WHO quickly responded with the co-ordination of the production of the diagnostic, guiding patient monitoring, collection of samples, and care. Some countries in the region as well as in the USA are monitoring Wuhan travelers for a fever to detect cases of 2019-nCoV before the virus continues to spread. Chinese, Thailand, Korean, and Japanese reports suggest that 2019-nCoV disease appears relatively mild compared to SARS and MERS. Recent trends in the Internet of Things (IoT) have seen a significant shift toward healthcare technology [28-32].

IoT advancements are progressively getting increasingly mentioned in human services with regard to improvement, testing, and preliminaries, with the expectation to be utilized as a piece of the two centers and homes. This Special Issue centers around best in class IoT advances, remote body

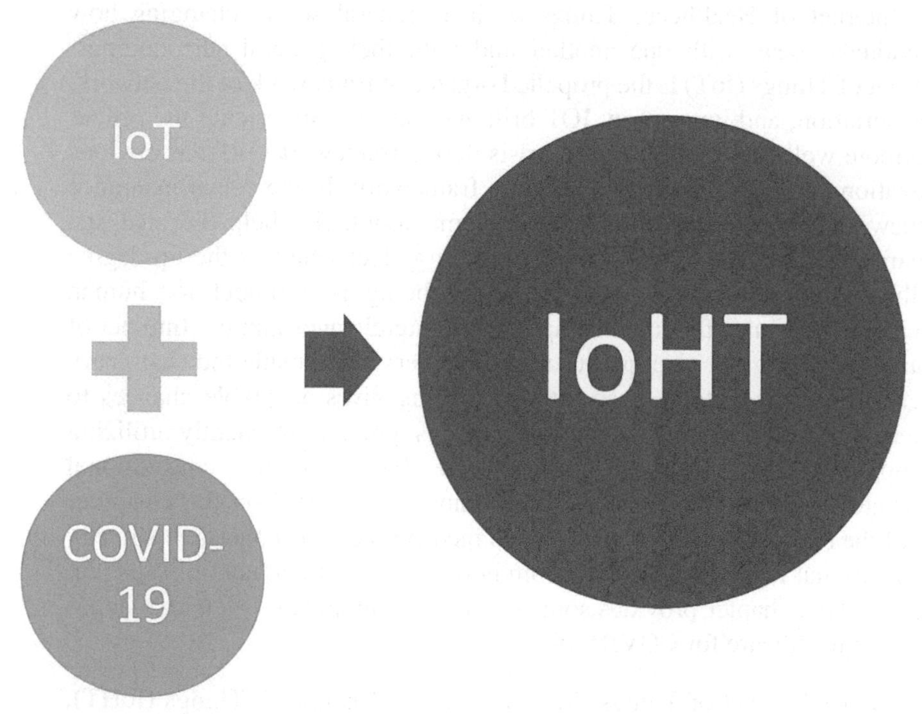

Figure 8.1 IoHT for COVID-19.

zone sensor systems, signal preparation and investigation, clinical imaging, and progressed inescapable medicinal services frameworks being utilized to screen explicit sicknesses/issues of patients. When all is said and done, numerous database groups and extra assets are required to store large information. Be that as it may, capacity and recovery are, by all account, not the only issues. Important examples are difficult to get from enormous information, for example, that relating to tolerant analytic data, which is likewise a basic issue. By and by, various rising applications are being created for different conditions. Sensors are regularly utilized in basic applications for on-going or the not-so-distant future [18-21]. Specifically, the Internet of Healthcare Things (IoHT) utilizes an accelerometer sensor, visual sensor, temperature sensor, carbon dioxide sensor, ECG/EEG/EMG sensor, pressure sensor, gyrator sensor, blood oxygen immersion sensor, moistness sensor, a breath sensor, and pulse sensor to watch and consistently screen patients' wellbeing. By wisely exploring and gathering a lot of clinical information (i.e., huge information), IoHT can improve the dynamic procedure and early illness analysis. Thus, there is a requirement for versatile artificial intelligence (AI) and insightful calculations that lead to progressively interoperable arrangements and that can settle on compelling choices in rising IoHT [17]. The Internet of Healthcare Things, IoHT to put it plainly, is an idea that depicts particularly recognizable gadgets associated with the Internet and ready to speak with one another, utilized in the clinical territory. These arrangements empower, for instance, restriction and continuous data about resources. Remote or programmed, the executives of assets are conceivable as well. This leads not exclusively to greater consideration and time investment funds, yet, additionally guarantees tolerant wellbeing [22-25]. On account of the IoT, overseeing clinical offices is increasingly effective with continuous access to gear, information, and patients' data. The most significant highlights of most IoHT arrangements are adaptability and personalization. Frameworks can be adjusted to each office, regardless of specialization or size. They can likewise be incorporated with existing advances and programming, associated and changed [15-16].

8.2 Related Work

Various scientists have proposed different models for IoT in healthcare and the forecast of different sorts of illnesses utilizing different procedures. This part centers on the work done in a similar zone. Ahn *et al.* [1] executed a framework for estimating the physiological signals in sitting position, for example, electrocardiogram (ECG) and ballistocardiogram (BCG) by utilizing a keen seat that detects the non-obliged bio-flags and can be observed utilizing a checking framework, for example, the one they had created giving an exemplary case of the use of IoT in medicinal services. Almotiri *et al.* [2] proposed an arrangement of m-wellbeing that utilizes cell phones to gather on-going information from patients in and store it on organizing servers associated with the web empowering access just to certain particular customers. Barger *et al.* [3] made a brilliant house office utilizing a sensor system to screen and track the developments of the patient in home and a model of the equivalent is additionally being tried. Chiuchisan *et al.* [4] proposed a structure to forestall the dangers to the patient in brilliant ICUs. The proposed framework suggests the patient's family members and specialists about any irregularity in their wellbeing status or their body developments and about the environment of the room with the goal that the vital careful steps can be taken. Dwivedi *et al.* [5] built up a structure to make sure about the clinical data that must be transmitted over the web for electronic patient record (EPR) frameworks in which they propose a multi-layered medicinal services data framework system which is a blend of public key infrastructure, smartcard, and biometrics innovations. Gupta *et al.* [6] proposed a model which measures and records ECG and other imperative wellbeing parameters of the patient utilizing Raspberry Pi and can be of incredible use for the emergency clinics and patients just as their relatives. Gupta *et al.* [7] presented a methodology utilizing Intel Galileo improvement load up that gathers the different information and transfers it to the database from where it tends to be utilized by the specialists and lessen the torment conceived by the patients to visit the medical clinic every single time to check their wellbeing parameters. Lopes *et al.* [8] proposed a structure dependent on IoT for impaired individuals to study and discover the IoT advancements in the medicinal services section that can profit them and their locale. They took two use cases to consider the most recent IoT innovations and its application that can be utilized primarily for impaired individuals. Nagavelli and Rao [9] proposed a novel strategy to foresee the seriousness of the disorder from the patient's clinical record utilizing a mining-based measurable methodology which they said as the level of sickness likelihood edge. Furthermore, to meet their objective, they have patched up an0 calculation that is generally expected to determine the hyperlink weight of the sites. Sahoo *et al.* [10] considered the human services the executive's framework and about the huge measure of patient information that is created from different reports. They further broke down the wellbeing parameters to anticipate the future wellbeing states of the patient or the said subject. They utilize cloud-based huge information expository stage to accomplish a similar utilizing the methods for likelihood. Tyagi *et al.* [11] investigated the job of IoT in human services and contemplated its specialized viewpoints to make it a reality and recognize the open doors for which they propose a cloud-based applied system in which the patients' clinical information and data can be safely moved, with the consent of the patient and their family by building a system among tolerant, emergency clinic, specialists, labs, and so forth. Xu *et al.* [12] introduced an information model to record and utilize the IoT information. Harleen *et al.* portrayed to check and comprehend IoHT programs [13]. Peruse *et al.* talked about cases emerging across China and different nations and areas [14].

8.3 IoT with Architecture

The expression "The Internet of Things" was authored by Kevin Ashton in an introduction to Proctor and Gamble in 1999. He is a prime supporter of MIT's Auto-ID Lab.

The "Thing" in IoT can be any device with any sort of implicit sensors with the capacity to gather and move information over a system without manual mediation. The inserted innovation in the article causes them to associate with interior states and the outer condition, which, thus, helps in choices making process. IoT is a system of physical nodes or individuals called "things" that are inserted with programming, hardware, system, and sensors that permit these items to gather and trade information. The objective of IoT is to stretch out to web availability from standard gadgets like PC, versatile, tablet to generally imbecilic gadgets like a toaster [1-4]. IoT makes practically everything "shrewd," by improving parts of our existence with the intensity of information assortment, AI calculations, and systems. The thing in IoT can likewise be an individual with a diabetes screen embed, a creature with GPS beacons, and so forth. The working of IoT is distinctive for various IoT reverberation frameworks (engineering). Nonetheless, the key idea of their working is comparable. The whole working procedure of IoT begins with the gadget themselves, for example, cell phones, advanced watches, electronic apparatuses, etc., which safely speak with the IoT stage and IoT architecture shown in Figure 8.2. The stages gather and break down the information from every one of the different gadgets and stages and move the most important information with applications to gadgets.

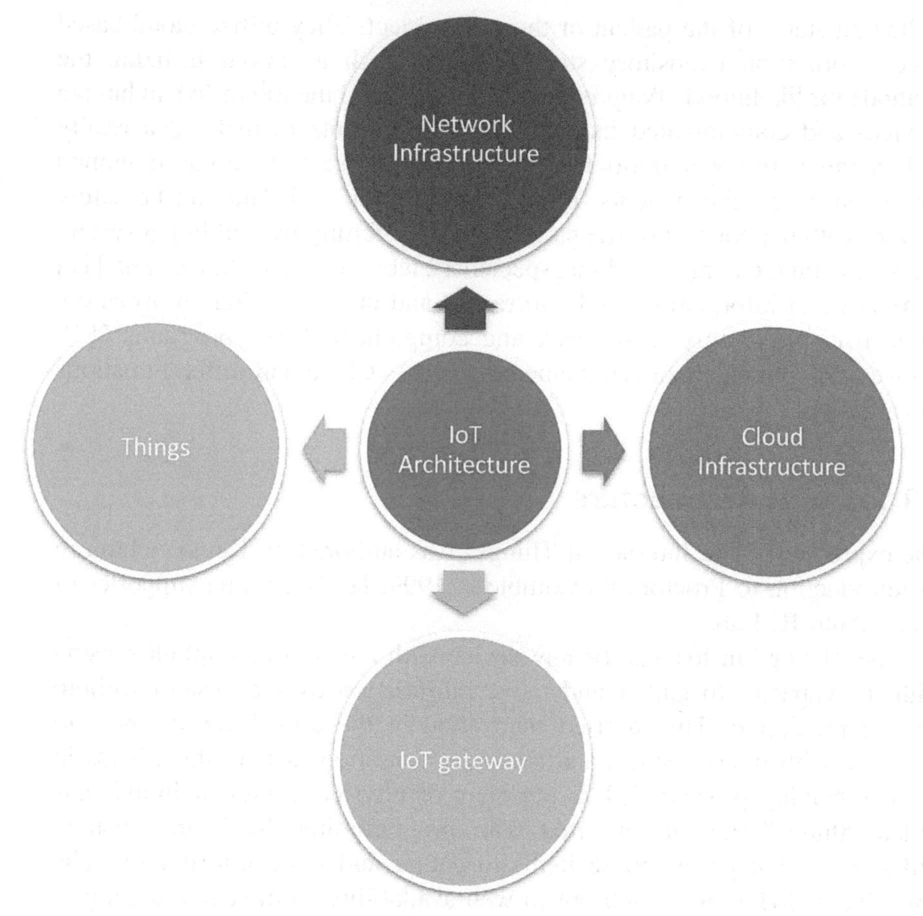

Figure 8.2 IoT architecture.

8.4 IoHT Security Requirements and Challenges

IoHT is prospering quickly in the clinical part and soon it will be across the board of this idea in every aspect of the clinical practice. Alongside these turns of events, there are likewise raising worries about a heap of security issues and difficulties. Right now, talk about a portion of these significant perspectives concerning security in IoHT empowered gadgets and systems. Since the improvement of IoHT gadgets and applications are at the level of the early stage, there are numerous constraints by and by that assume an imperative job at the beginning of difficulties looked by IoHT. These constraints incorporate vitality confinements, memory restrictions, and computational force impediments. With the progressing mechanical advancements in the equipment, these impediments are diminishing with the progression of time, and gadgets are consistently improved with better battery life, expanded memory sizes, and higher handling speeds [13].

Different difficulties, which are more on the product side, incorporate privacy of data, the respectability of delicate clinical information, accessibility of administration, and adaptation to non-critical failure, confirmation, and approval. Secrecy of data forestalls the entrance of clinical data to unapproved clients. For this, few strategies have been actualized and more are required to be created as the innovation keeps on upgrading. The honesty of information forestalls any change of information during its travel arrange by any enemy sniffing the IoHT organize. Another significant angle with regard to IoHT challenges is the accessibility of administration shown in Figure 8.3. It guarantees that the human services administrations will be

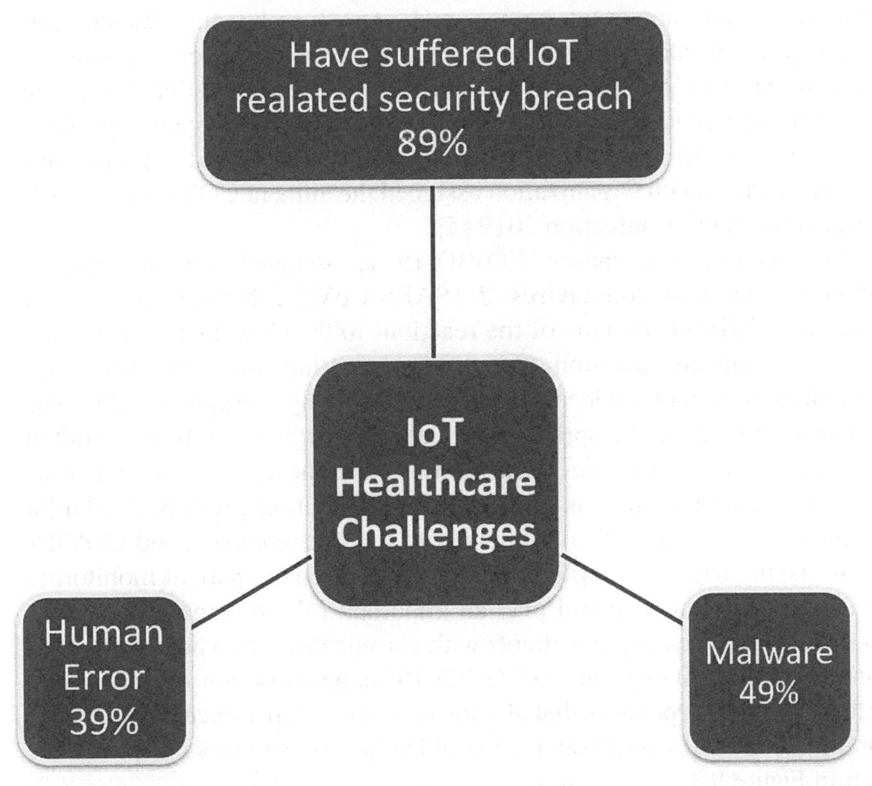

Figure 8.3 IoHT healthcare challenges.

accessible to the end-clients when and where required. Any vacation and forswearing of administration would demoralize the embracement of IoHT administrations and applications. Also, these administrations ought to be an issue open-minded. This desires the administrations to be as yet accessible even in the hours of deficiencies and assaults. In conclusion, one of the most significant moves identified with security is the component for guaranteeing validation and approval. Validation guarantees that the IoHT gadget is conveying to the friend it should impart and there is no bogus nearness. The approval guarantees that solitary approved hubs are permitted to use and access IoHT to organize friends and assets. For additional subtleties on these difficulties, per user is alluded to which present a point-by-point conversation on them [13].

8.5 COVID-19 (Coronavirus Disease 2019)

Coronaviruses are important human and animal pathogens. At the end of 2019, a novel coronavirus was identified as the cause of a cluster of pneumonia cases in Wuhan, a city in the Hubei Province of China. It quickly spread, bringing about a scourge all through China, trailed by an expanding number of cases in different nations all through the world. In February 2020, the World Health Organization assigned the ailment COVID-19, which represents coronavirus infection 2019 [5].

The infection that causes COVID-19 is assigned extreme intense respiratory condition coronavirus 2 (SARS-CoV-2); beforehand, it was alluded to as 2019-nCoV. One of the reactions to the COVID-19 emergency has been a sincere exertion at social separating that has numerous representatives currently telecommuting and avoiding workplaces [26]. The objective is to forestall the spread of the infection, which can be transmitted via air just by contact or on surfaces. Basic tele-health services (such as remote consultation), which are not strictly IoT solutions, will be used a lot more during this crisis. IoT could help with future pandemics, and COVID-19 could be the trigger to explore new solutions. Remote patient monitoring (RPM) and telemedicine could play an important role in managing a future pandemic [27]. For example, patients with chronic diseases who have to self-isolate to reduce their exposure to COVID-19 but need continuous care would benefit from RPM. Operators that already have some experience in RPM such as Orange, Telefónica, and Vodafone could help. Corona virus symptoms are shown in Figure 8.4.

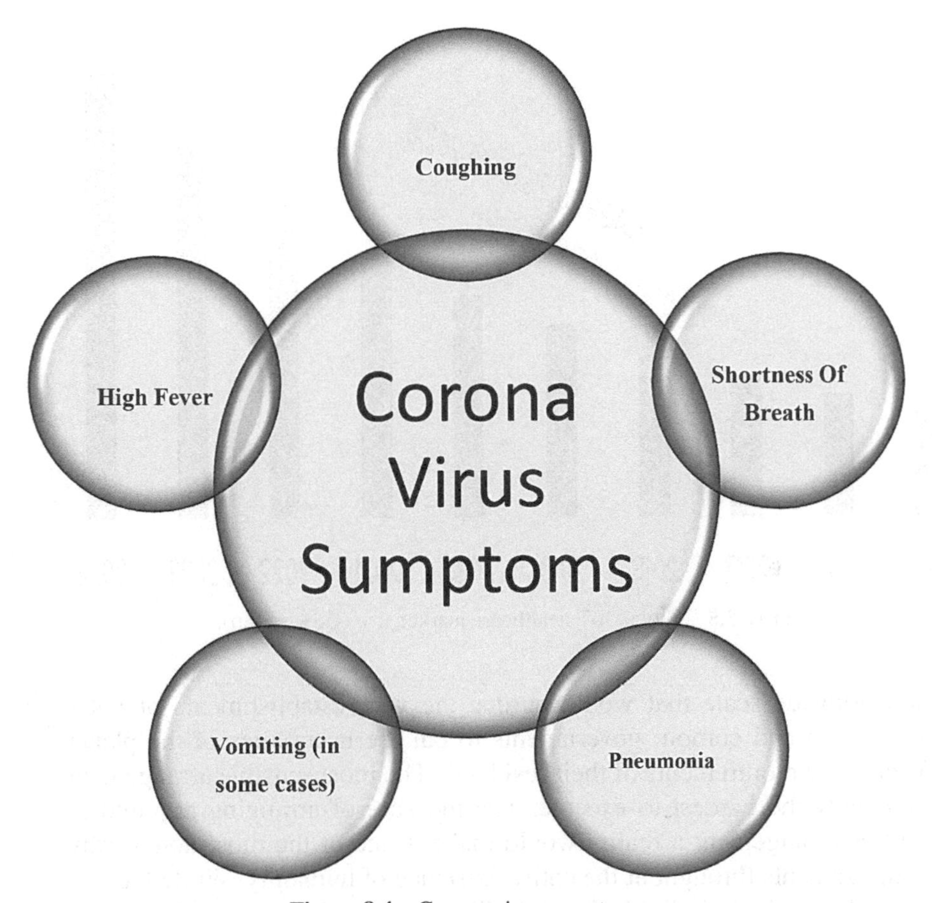

Figure 8.4 Corona virus symptoms.

8.6 The Potential of IoHT in COVID-19 Like Disease Control

The IoT, a system of interconnected frameworks and advances in information examination, AI, and universal network can help by giving an early admonition framework to control the spread of irresistible illnesses like coronavirus. In the Asia-Pacific area, China drives the path in IoT reception followed by Japan; the nation is foreseen to burn through $20.2 billion in IoT by 2024 as shown in Figure 8.5.

The basic answer may be for endeavors, urban communities, and national governments to, by and large, make a huge worldwide system of sensors to identify infections. In any case, this would require arranging and usage

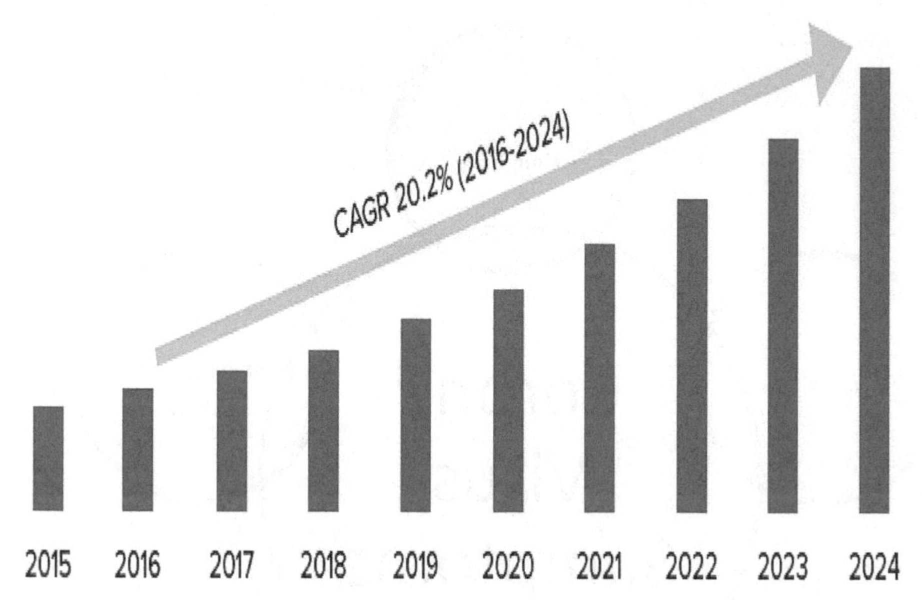

Figure 8.5 Global IoT healthcare market size (US$ billion).

on a worldwide scale that would burden the very establishments of vote-based system and commit governments to put the necessities of the planet in front of the requirements of their residents. The most sensible arrangement is frequently the hardest to execute. The measure of arranging required to make this arrangement a reality would make it one of the most noteworthy accomplishments throughout the entire existence of humanity. While I am an inalienably idealistic individual (while likewise practical), I see this as the "sacred goal" of IoT opportunity in the long haul.

So what is a portion of the down to earth ways that IoT can help in the close to term? The initial phase in irresistible illness control is located. While a worldwide system of sensors is probably not going to occur within a reasonable time frame, China can actualize such a system in the nation. China has a background marked by executing wide-territory IoT arrangements (for example, video observation) on a scale that has never been seen. So why not a system of infection identification sensors? Couple that with facial acknowledgment and area, existing reconnaissance cameras to distinguish, follow, and screen individuals that may have gotten the coronavirus. An additional layer has likewise tracked each person that a tainted patient reached. While this may seem like a police state to some, at last, utilizing IoT and AI might be the most sensible approach to keep profoundly irresistible illnesses from spreading quickly in a world that is getting littler consistently with air travel.

8.7 The Current Applications of IoHT During COVID-19

Right now, IoHT is used to deal with certain parts of the COVID-19. For instance, rambles are, as of now, utilized for open reconnaissance to guarantee isolation and the wearing of veils. Computer-based intelligence has likewise been utilized to foresee future flare-up regions as shown in Figure 8.6.

8.7.1 Using IoHT to Dissect an Outbreak

With the various and different datasets gathered by cell phones, IoHT can have a lot more applications during a scourge. IoHT can be utilized to follow the inception of an episode. An ongoing report by scientists at MIT utilized collected cell phone information to follow, in granular subtleties of short separations and periods, the spread of dengue infection in Singapore during 2013 and 2014. Subsequently, overlaying geographic data (GIS) on IoHT

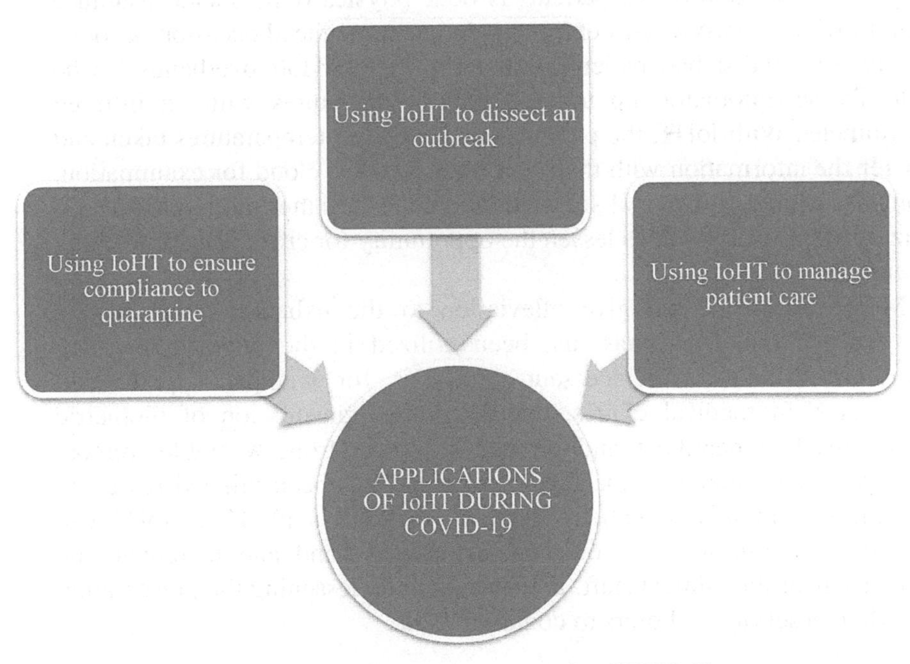

Figure 8.6 Applications of IoHT during COVID-19.

versatile information from contaminated patients can complete two things. Upstream, which can help disease transmission specialists as they continued looking for quiet zero; downstream, which can help distinguish all the people who have come into contact with the contaminated patients and may, in this way, likewise be tainted

8.7.2 Using IoHT to Ensure Compliance to Quarantine

IoHT can likewise be utilized to guarantee tolerant consistency once the possibly tainted people go into isolation. General wellbeing staff can screen which patients stayed isolated, and which patients have penetrated the isolation. The IoHT information will likewise assist them with finding who else might be presented because of the break.

8.7.3 Using IoHT to Manage Patient Care

The adaptability of IoHT likewise proves to be useful for checking all the patients who has high sufficient chance to warrant isolation, however, not genuine enough to warrant in-emergency clinic care. At present, the everyday registration of the patients is done physically by social insurance laborers who go entryway-to-entryway. In one announced occasion, a social insurance specialist had patients remaining in their loft overhangs, so he could fly an automaton up to take their temperatures with an infrared thermometer. With IoHT, the patients can have their temperatures taken and transfer the information with their cell phones to the cloud for examination. Along these lines, medicinal services laborers can gather more information utilizing less time as well as lessen the opportunity for cross-disease with the patients.

Moreover, IoHT can give alleviation to the exhausted staff at the emergency clinic. IoHT has just been utilized in the remote observing of in-home patients with incessant conditions, for example, hypertension or diabetes. In medical clinics, telemetry, the transmission of biometric estimations like heartbeat and circulatory strain from wearable, remote instruments on patients to the focal observing, has been utilized to screen an enormous number of patients with insignificant staff. Here, IoHT can be utilized to diminish the outstanding task at hand and increment the productivity of the clinical staff, at the same time, lessening the presentation of medicinal services laborers to contamination.

8.8 IoHT Development for COVID-19

While the world continues to live under the cloud of uncertainties and fear arising because of the COVID-19 outbreak, its impact on everyone's psychology and behavior is unimaginable. Simple, yet most effective, protective measures against the novel coronavirus such as social distancing and regular hand-washing, along with protecting us from the spread are also shifting our behaviors slowly. Everyone is trying to cut down contact with strangers or with items/exposed surfaces that are touched by several people to avoid the possibility of catching the infection. There are few items such as switches and remote controls that everyone ends up touching multiple times in a day as shown in Figure 8.7, even amid the healthcare we are currently witnessing.

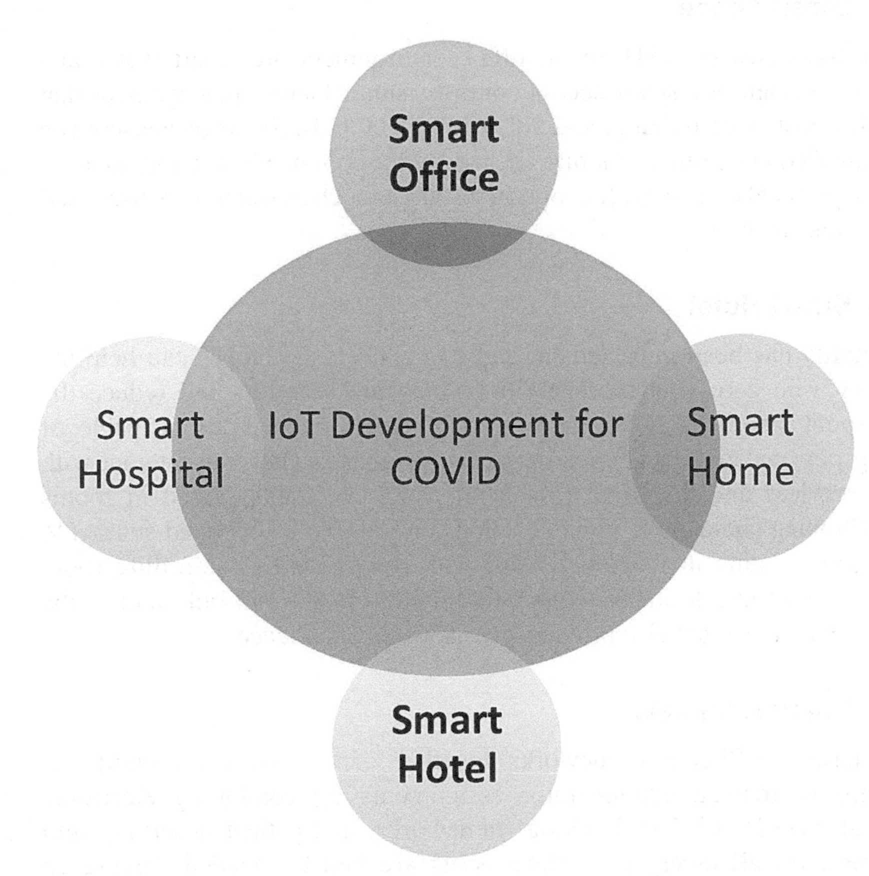

Figure 8.7 IoHT developments for COVID-19.

8.8.1 Smart Home

There are a plethora of IoHT devices such as smart video door phones, smart locks, smart lights, smart TVs, and smart air conditioners (ACs), among others, that can add a lot of convenience for smart home-use cases. These devices can be controlled toothlessly by use of voice assistants such as Google Assistant, Alexa, and Siri or through your mobile phone. For example, video door phones can help users to check visitors at their home. With two-way audio of the video door phone, users can interact with visitors through mobile apps and can eliminate undesired and less-important visitors [22]. Particularly, it can be very useful when you are dealing with food and e-commerce deliveries.

8.8.2 Smart Office

Major use cases of IoHT in an office environment are smart water and beverage dispensers, smart access controls, smart facility management that includes control of lighting, and AC using IoHT [23]. Smart dispensers can be controlled via gestures, mobile app, and voice commands, leaving no room for any physical contact with common resources such as water dispensers and coffee machines.

8.8.3 Smart Hotel

Hospitality has been impacted the most by COVID-19. IoHT can help the industry with zero contact check-ins, which are seamless and reduce the time spent by guests at the hotel's lobby. IoHT tech is also capable of adding personalized experience in rooms for guests. They can interact with room services and the front desk using voice assistants placed in rooms [24]. Flipping channels or movie catalogs on your voice-operated smart TV, operating curtains in the room using voice commands, controlling room lights, and setting desirable temperature/mode on ACs not only add to the guest's experience but also provides a touch-less experience.

8.8.4 Smart Hospitals

Associating wellbeing frameworks together can decrease an enormous measure of manual administrator assignments by combining electronic medical records (EMRs), booking frameworks, and patient observing into one spot. As all emergency clinic assets are being extended, having an instrument to screen patients all around the emergency clinic and guarantee that medicine is conveyed viably will be an enormous assistance. Aside from the numerous instances of coronavirus separating off pieces of emergency clinics and taking up a tremendous extent of medical clinic staff's time and consideration, other high-hazard patients despite everything require similar degrees of care. Gadgets that screen glucose levels for diabetic patients monitor circulatory strain and pulse levels, and caution to issues, for instance, can permit emergency clinic staff to deal with these patients remotely while in another segment of the medical clinic [25].

8.9 Conclusion

The basic innovation and the IoHT parts that can be utilized to empower a social insurance framework to manage illness flare-ups as of now exist; in any case, they are divided and not yet associated. Hence, the framework should have the option to develop its foundation rapidly to interface the parts of information assortment, preparing, and capacity so that the system can scale and expand for disease tracking, preventive quarantine, and the in-patient care of the infected. And this chapter only provides challenges and developments of IoHT in healthcare for COVID-19 and research could be extended further such that big data, augmented reality, and cognitive systems should be considered in future researches.

References

[1] B. G. Ahn, Y. H. Noh, and D. U. Jeong. Smart chair based on a multi heart rate detection system. In 2015 IEEE SENSORS, pages 1–4, Nov 2015.

[2] B. Bhushan, C. Sahoo, P. Sinha, and A. Khamparia. Unification of Blockchain and Internet of Things (BIoT): Requirements, working model, challenges and future directions. Wireless Networks. 2020. doi:10.1007/s11276-020-02445-6

[3] T. S. Barger, D. E. Brown, and M. Alwan. Health-status monitoring through analysis of behavioral patterns. IEEE Transactions on Systems, Man, and Cybernetics-Part A: Systems and Humans, 5(1):22–27, Jan 2005. ISSN 1083-4427.

[4] N. Sharma, I. Kaushik, V. K. Agarwal, B. Bhushan, and A. Khamparia. Attacks and security measures in wireless sensor network.

Intelligent Data Analytics for Terror Threat Prediction, 237-268. 2021. doi:10.1002/9781119711629.ch12

[5] A. Dwivedi, R. K. Bali, M. A. Belsis, R. N. G. Naguib,P. Every, and N. S. Nassar. Towards a practical healthcare information security model for healthcare institutions. In 4th International IEEE EMBS Special Topic Conference on Information Technology Applications in Biomedicine, 2003., pages 114–117, April 2003.

[6] M. S. D. Gupta, V. Patchava, and V. Menezes. Healthcare based on IoHT using raspberry pi. In 2015 International Conference on Green Computing and Internet of Things (ICGCIoHT), pages 796–799, Oct 2015.

[7] P. Gupta, D. Agrawal, J. Chhabra, and P. K. Dhir. IoHT based smart healthcare kit. In 2016 International Con- ference on Computational Techniques in Information and Communication Technologies (ICCTICT), pages 237–242, March 2016.

[8] N. V. Lopes, F. Pinto, P. Furtado, and J. Silva. IoHT architecture proposal for disabled people. In 2014 IEEE 10th International Conference on Wireless and Mobile Computing, Networking and Communications (WiMob), pages 152–158, Oct 2014.

[9] R. Nagavelli and C. V. Guru Rao. Degree of disease possibility (ddp): A mining based statistical measuring approach for disease prediction in health care data min- ing. In International Conference on Recent Advances and Innovations in Engineering (ICRAIE-2014), pages 1–6, May 2014.

[10] P. K. Sahoo, S. K. Mohapatra, and S. L. Wu. Analyzing healthcare big data with prediction for future health condition. IEEE Access, 4:9786–9799, 2016. ISSN 2169-3536.

[11] S. Tyagi, A. Agarwal, and P. Maheshwari. A conceptual framework for IoHT-based healthcare system using cloud computing. In 2016 6th International Conference - Cloud System and Big Data Engineering (Confluence), pages 503–507, Jan 2016.

[12] B. Xu, L. D. Xu, H. Cai, C. Xie, J. Hu, and F. Bu. Ubiq- uitous data accessing method in IoHT-based information system for emergency medical services. IEEE Transac- tions on Industrial Informatics, 10(2):1578–1586, May 2014. ISSN 1551-3203.

[13] HKaur, M, Atif, R, Chauhan, an Internet of Healthcare Things (IoHT)-Based Healthcare Monitoring System. In: Mohanty M., Das S. (eds) Advances in Intelligent Computing and Communication. Lecture Notes in Networks and Systems, vol 109. Springer, Singapore, 2020.

[14] J. Read, J. Bridgen, D. R. Cummings, and C. P. Jewell. Novel coronavirus 2019-nCoV: early estimation of epidemiological parameters and epidemic predictions. medRxiv.2020.

[15] A. Kumar, A. O. Salau, S. Gupta, and K. Paliwal. Recent trends in IoT and its requisition with IoT built engineering: A review. Advances in Signal Processing and Communication, 2019. pp. 15-25.

[16] S. Sharma, A. Kumar. Enhanced Energy-efficient heterogeneous Routing Protocols in WSNs for IoT Application, IJEAT, ISSN: 2249-8958VOL. 9 ISSUE-1, 2019.

[17] A. K. Rana, R. Krishna, S. Dhwan, S. Sharma, and R. Gupta. Review on Artificial Intelligence with Internet of Things-Problems, Challenges and Opportunities. In *2019 2nd International Conference on Power Energy, Environment and Intelligent Control (PEEIC)*, pp. 383-387, October 2019. IEEE.

[18] A. A. Adamu, D. Wang, A. O. Salau, and O. Ajayi. An Integrated IoT System pathway for Smart Cities. International Journal on Emerging Technologies, 11(1), 01–09 (2020).

[19] Y. M. Mekki, M. M. Mekki, M. A. Hammami and S. M. Zughaier, "Virtual Reality Module Depicting Catheter-Associated Urinary Tract Infection as Educational Tool to Reduce Antibiotic Resistant Hospital-Acquired Bacterial Infections," 2020 IEEE International Conference on Informatics, IoT, and Enabling Technologies (ICIoT), Doha, Qatar, 2020, pp. 544-548, doi: 10.1109/ICIoT48696.2020.9089488.

[20] A. Kumar and S. Sharma. IFTTT Rely Based a Semantic Web Approach to Simplifying Trigger-Action Programming for End-User Application with IoT Applications. Semantic IoT: Theory and Applications: Interoperability, Provenance and Beyond, p. 385.

[21] C. Chokphukhiao, R. Patramanon, K. Maitree and S. Kasemvilas, "Health Monitoring Platform for Emergency Medicine: User Perspective and Implementation," 2020 Joint International Conference on Digital Arts, Media and Technology with ECTI Northern Section Conference on Electrical, Electronics, Computer and Telecommunications Engineering (ECTI DAMT & NCON), Pattaya, Thailand, 2020, pp. 133-136, doi: 10.1109/ECTIDAMTNCON48261.2020.9090731.

[22] B. Bhushan and G. Sahoo. Recent Advances in Attacks, Technical Challenges, Vulnerabilities and Their Countermeasures in Wireless Sensor Networks. Wireless Personal Communications, 98(2), 2037-2077. 2017. DOI: 10.1007/s11277-017-4962-0

[23] K. Vijayakumar and V. Bhuvaneswari, "A Ubiquitous first look of IoT Framework for Healthcare Applications," 2020 International Conference on Emerging Trends in Information Technology and Engineering (ic-ETITE), Vellore, India, 2020, pp. 1-7, doi: 10.1109/ic-ETITE47903.2020.146.

[24] A. Kumar and S. Sharma. Internet of Things (IoT) with Energy Sector-Challenges and Development. In Electrical and Electronic Devices, Circuits and Materials. pp. 183-196, 2021. CRC Press.

[25] A. D. Acharya and S. N. Patil, "IoT based Health Care Monitoring Kit," 2020 Fourth International Conference on Computing Methodologies and Communication (ICCMC), Erode, India, 2020, pp. 363-368, doi: 10.1109/ICCMC48092.2020.ICCMC-00068.

[26] H. Kang et al., "Diagnosis of Coronavirus Disease 2019 (COVID-19) with Structured Latent Multi-View Representation Learning," in IEEE Transactions on Medical Imaging, doi: 10.1109/TMI.2020.2992546.

[27] L. Li et al., "Characterizing the Propagation of Situational Information in Social Media During COVID-19 Epidemic: A Case Study on Weibo," in IEEE Transactions on Computational Social Systems, vol. 7, no. 2, pp. 556-562, April 2020, doi: 10.1109/TCSS.2020.2980007.

[28] D. Dong et al., "The role of imaging in the detection and management of COVID-19: a review," in IEEE Reviews in Biomedical Engineering, doi: 10.1109/RBME.2020.2990959.

[29] Rana, Arun Kumar, and Sharad Sharma. "Industry 4.0 Manufacturing Based on IoT, Cloud Computing, and Big Data: Manufacturing Purpose Scenario." In *Advances in Communication and Computational Technology*, pp. 1109-1119. Springer, Singapore, 2021.

[30] Rana, Arun Kumar, and Sharad Sharma. "Contiki Cooja Security Solution (CCSS) with IPv6 Routing Protocol for Low-Power and Lossy Networks (RPL) in Internet of Things Applications." In *Mobile Radio Communications and 5G Networks*, pp. 251-259. Springer, Singapore, 2021.

[31] A. K. Rana and S. Sharma. The Fusion of Blockchain and IoT Technologies with Industry 4.0. In Intelligent Communication and Automation Systems. pp. 275-290, 2021. CRC Press.

[32] Kumar, Arun, and Sharad Sharma. "Demur and Routing Protocols With application in Underwater Wireless Sensor Networks for Smart City." In *Energy-Efficient Underwater Wireless Communications and Networking*, pp. 262-278. IGI Global, 2021.

9

An Integrated Approach of Blockchain & Big Data in Health Care Sector

Nitin Tyagi[1], Bharat Bhushan[2], Siddharth Gautam[3], Nikhil Sharma[3*] and Santosh Kumar[4]

[1]HMR institute of Technology & Management, Delhi
[2]School of Engineering and Technology (SET), Sharda University, India
[3]HMR Institute of Technology & Management, Delhi
[4]ITER-SOA, Bhubaneswar, Odisha
E-mail: nitintyagi5631@gmail.com, bharat_bhushan1989@yahoo.com, siddharthinfo92@gmail.com, nikhilsharma1694@gmail.com, santoshkumar@soa.ac.in

Abstract

Blockchain has many inherent features like decentralized storage, authentication, distributed ledger, immutability and safety, which are more practical in the fields of healthcare. Blockchain applications in the medical sector normally require stronger authentication, sharing of records and interoperability due to stringent legislative requirements, including the 1996 HIPAA Act. Researchers from both the academic and the industry have begun to explore applications ready for use in healthcare, using current technology. These applications include detection of fraud, intelligent contracts and identity validation. In addition to these reforms, technology blockchain has concerns about its particular weaknesses and issues such as core management, mining attacks and incentives. This chapter discusses an overview of the blockchain that includes the Health Care Big Data, as well as blockchain solutions for big data on healthcare. In this chapter, a holistic model of the healthcare system, specifically for developing a smart healthcare system, has been proposed.

Keywords: Medrec, PSN, Big Data, Blockchain, Cyber Security, Fighting Drugs.

9.1 Introduction

Blockchain technology was implemented in 2008 with outdated features such as authentication, decentralization, and immutability. A successful use of bitcoin has given rise to debate and ideas to allow blockchain technical use in other data-driven fields like healthcare- approximately 400 million full-scale transactions (March 19, 2019) [1]. Initially, blockchain was created for the alternative of currency. A decentralized and stable currency which can be used all over the world as a means of trade. With the assistance of blockchain technology, one can send electronic payments directly from one party to another without any interference by third-party vendors. Network can add another protection layer by avoiding data breach in transactions operating time stamp [2]. This property makes blockchain better than other technologies and gives confidence, transparency, and accessibility to any exchanges in different industries. A few years after bitcoin 's release, codes were published as open-source allowed researchers, leading to several different blockchain-based applications and protocols being developed. Scientists started to realize the potential of blockchain and brought this technology beyond the realm of non-financial industries, as well. Blockchain is a decentralized, peer-to-peer distributed laser that can record all transactions that take place on a network. That makes it useful for any property, data, currency, details, etc. exchange [3]. Recent advances in this technology have occurred aimed at implementing non-financial blockchain applications. This technology is therefore also expanded in other industries, such as human resources, identification management, supply chain management, and so on. Health services are one of the most widely used areas of activity in different industries, with the growing enthusiasm for blockchain across various industries. Practitioners in the health care field have the burden of fragmented and delayed data, prolonged and delayed communications, and keep patient records. The architecture based on blockchain is also introduced to gather practical information through technology embedded in the framework [4]. The design of the blockchain has changed over time. Consequently, several types of blockchain were created. In some implementations, any node can join the network freely [5]. This form of implementing blockchain is called a public blockchain. There's another implementation in which any node that wants to join the blockchain network needs to be approved and allowed to join the network, that's called permitted blockchain. And in a legal blockchain, only a few nodes can become miners [6]. The prohibited chain is therefore tiny and fairly quick and stable. When only one node is permitted to be a miner, then it is a private blockchain, but private blockchain loses most of the property of decentralization as power is only granted to one node. If more than one node operates as a miner it becomes a blockchain of the Consortium [7]. Blockchain 's fundamental property lies in its IT (Information Technology) architecture and the sequence of data entries that enable unbroken safe and open transactions [8]. The major contribution of this chapter is to enable the healthcare center to maintain privacy and security among its stakeholders. Yet, discrete blockchain applications of the various functionalities of the healthcare system will give the added complexity to the current system. In this chapter, a holistic model of the healthcare system, specifically for developing a smart healthcare system, has been proposed. Blockchain technology can also enhance interoperability between different systems that are critical for improving the existing health care system. Healthcare blockchain can replace already outdated IT systems with a new interoperable system [9]. The chain, in addition to providing interoperability, impedes the benefits of cryptographically safe and irreversible transactions and thus ensures privacy among the parties involved.

The rest of the paper is organized as follows. Section 2 presents the role of blockchain technology in health care. Section 3 presents the overview of Blockchain & Big Data in health care. In Section 4, various types of application of Big Data for Blockchain are discussed. Section 5 explains about the integration of blockchain with Big Data and its solutions followed by the conclusion in Section 6.

9.2 Blockchain for Health care

The health care sector has introduced plenty of incentives for adopting blockchain. It includes many stakeholders such as the patient, hospital, physician, clinical researcher, a few policy insurers names and drug suppliers. There's a lot of data sharing between various industry stakeholders. It then becomes critical that the documents are kept illegally. An agency cannot manipulate the data by interfering with records or through any other means.

The most visual data from any stakeholders may impact patient care and pose a serious life threat [10]. The health industry is facing a major problem with storage and data sharing which threatens data privacy and protection. Researchers have created a blockchain network, called blocHIE [11], which stores and shares information between different stakeholders involved in the process to ensure that data does not change. Besides the storage and sharing of information, the storage of daily data in the healthcare sector is also a significant factor. Therefore, the blockchain can manage the whole healthcare data network, where all stakeholders can securely access data from a single point [12]. Blockchain may have a larger effect on clinical healthcare research, as it enables data to be processed, exchanged, and tracked. This blockchain, due to numerous scams in recent years, has improved the reliability of commonly crafted diagnostic testing. This technology will serve as a constructive catalyst for better practice for clinical research and be a step closer to greater network openness and trust. It can lead to improved secure contact between research and communities affected and enhance trust in research communities [13]. Nowadays, Blockchain has several applications, as it has tremendous potential to affect healthcare technology.

9.2.1 Healthcare data sharing through gem Network

The healthcare industry has massive quantities of data, records, and other sensitive documents that need to be maintained and stored in centralized repositories, and there is always a challenge to its protection and interoperability [14]. There are also some problems involved in exchanging confidential records when a patient is moved from one hospital to another hospital, medical institutes, interstate or overseas, as each particular moment a patient meets another doctor, there is always a contact difference between the medical staff as a result of which new records are produced. Gem Health Network has developed a possible solution to all of the above-mentioned problems using Ethereum Blockchain Technology and thus building a decentralized network of healthcare professionals that can access all relevant information at the same time removing the constraint of centralized storage [15]. It also helps to provide real-time information, thus improving accountability, accessibility, and mitigating the risk of negligence in health problems that might arise from inaccurate information or misinformation [16]. Guard-time is a healthcare platform that uses the advanced blockchain technology [17]. Thus, Estonian citizens, healthcare workers, providers, and health insurance companies can access all information on medical treatment using this technology in the above-mentioned manner. The Gem Health Network seeks to build a bridge between doctors and patients so that they can share records [18].

9.2.2 OmniPHR

This model was created by three experts, namely Roehrs, Costa, and Righi, which allows patients to preserve their health records and offers an overview of the health records of patients held in various healthcare centers and providers [19]. Furthermore, this system is structured in such a way as to overcome the challenges faced by providers of healthcare services [20]. EHR stands for electronic health records (EHRs) and is maintained in compliance with government regulations and maintains a consistent exchange of information between the state and country lines, thus keeping data up-to-date and true to date. This information varies from PHR (Patient Health Data), as these records are kept secure by medical personnel without contact with patients, while PHR records are handled by patients [21]. In particular, OmniPHR serves as a practical model leveraging blockchain technology that focuses on transparency and interoperability, offering a full database of patient total health records that are processed and transmitted through various medical platforms [22]. Also, OmniPHR aims to challenge the architectural delivery and organizational problems that can be scalable, elastic in approach, and interoperable [23]. The PHR is stored in an ordered hierarchical manner that ends up encrypted by the specific data network's data blocks. Patients can also use this system anywhere, whether at home, in offices or in a hospital [24]. Providers can compose the PHR by means of devices capable of supplying data while consumers can use different devices to access the data anywhere. The model can be set up using an application server built as a network that includes routing from one peer to the other with different responsibilities [25]. Often there are several other components like data sources etc. First, data flows from the data sources and is then fed into an open standard framework and transformed into middleware afterward. The data sources are encrypted in the privacy module as well as authenticated in the protection system, and later the data is stored in repositories. OmniPHR laboratory data assesses the efficiency of the specific model of data sources. OmniPHR also uses openEHR is an open standard for supporting the hierarchical organization of data blocks. OmniPHR can be tested in a university environment, using simulated datasets [26]. The key results include managing data networks so that it raises the number of users without reducing the distribution time and segmenting the clusters in data blocks using chord algorithms such that the number of hops to access the records can be minimized. Thus, a stable model is obtained with reduced latency [27]. However, some limitations are also observed in which one of the main limitations is that each type of data set up by the network model must be followed by the standard and if any data does not follow the required standards then data cannot be shared and therefore data providers are responsible for verifying the entry of data such as patient diagnosis and demographics. The drawback is found that as the patient was ill and could not take care of that data, a patient has to give access to his data to someone else. OmniPHR does not consider key management and recovery problems in case of loss and leakage [28].

9.2.3 Medrec

Medrec was founded by four people, namely Azaria, Ekblaw, Vieira, and Lippman [29]. This is decentralized record management in nature, which handles EMRs using blockchain technology. In particular, medrec is used to maintain security, reliability through modular designs that incorporate local data storage with existing providers and provide patients and medical professionals with solutions. A collection of miners enable Blockchain transaction processes. Medrec opportunities data economists also provide patients and providers with options to report their data or information as metadata. It focuses on evaluating and applying the System methodology before any field study and consists of block content that gives data ownership and viewership permissions that can be shared via network routing (P2P network). Smart contracts allow automation, tracking transitions such as changes related to viewing rights, and any addition to existing records information or creating new records. Therefore, the information is kept safe and only the patient can allow data sharing, any changes or adding information to records between providers is a huge focus on security [30]. When any changes in details are suggested, an automatic notice will be sent to the patient allowing the patient to check the proposed changes in records and then the patient will agree to approve and deny the data accordingly. The aggregation of all patient-provider relationships is listed in a specified document and thus results in a single point of reference that makes it easier to search for any updates and improvements in medical records information. A cryptographic key for authentication is given to validate identity, and an established type of ID may be added to a social security number to ensure protection as employed in DNS-like techniques [31].

MedRec consists of different contracts such as registrar, description, and patient-provider relationship contracts, thereby helping to recognize a patient, maintain patient information and satisfy any other arrangement that the patient may have. Also, miners enable patient and provider nodes and are generated and authenticated using Ethereum blockchain technology. Thus, the model in its entirety reflects the advantages of a decentralized database model that smoothes the information sharing process between patients and providers without any possibility of failure and enhances the interoperability of sharing records. Additional advantages are that it seeks to address the data mining problems which function as a threat when applying blockchain implementation in the medical, healthcare domain [32]. Notwithstanding its impressive advantages, it has few drawbacks. The key drawbacks include maintaining an individual's information protection, being unable to handle or address digital rights problems, and being able to use data to conclude the patient-provider relationship. The high number of transactions is also an obstacle for blockchain technology [33]. The program needs permission to access information that can only be obtained by the patient because the surgeon cannot access the medical record without authorization in case of an emergency. Hence, the framework in today's advanced real world can be impractical. The system needs to be improvised by designing a two-tier access control system so that a trusted user (could be a former or family physician, medical practitioner, and health department) can access it in case of a life-threatening emergency [34].

9.2.4 PSN (Pervasive Social Network) System

A healthcare system will use PSN as a network that requires wireless sensing techniques for mobile computing. Yet there is a central drawback to fully practical ideas such as safe sharing of health data with other network nodes. Researchers are therefore designing a stable healthcare network based on PSN, using two protocols that involve an adapted variant of IEEE 802.15.6 and the blockchain technique respectively [35]. The first protocol handles the show of authentication by making ties that are protected with unstable computation. Therefore, connections that are secured between sensor nodes and mobile networks are developed in the WBAN region. The second protocol shares the health data between PSN nodes using blockchain and thus enables the sharing and access of health data from one PSN node to other PSN nodes by visiting each other network. The PSN program uses blockchain technology to promote data exchanged between the patient and the physicians or health care providers using the above two protocols. This can be demonstrated by using a practical example where a patient named Himani suffers from hypertension and consult with her doctor, Gautam. Now Gautam needs blood pressure information that is stored in her wearable blood pressure monitor [36]. Using the PSN technique, Gautam can access the required information by following these steps: Himani establishes a secure link between her smartphone and wearable wristband, and a blockchain transaction is sent from a miner node to one PSN node [37]. The new block is verified by the miner node by creating a blockchain generating signature.

9.2.5 Healthcare Data Gateway

The blockchain technology is based on HDG (i.e. Data Gateway for the Health Services) and is designed to facilitate patient information to monitor patient medicine records [38]. In other words, patient data can be controlled from one system to another and shared safely. Smartphone applications are supported by the so-called HDG technology. HDG helps support the latest 5 G network in smartphones in the fast networking also known as wireless networking. This application mainly consists of three layers: data management layer, storage layer, and data usage layer, all three layers have their properties and these layers perform various tasks. , For example, Data management layer interconnects HDG, storage layer provides space for health data in the system and ultimately data usage layer allows the user to access health department data.

9.2.6 Resources that are virtual

It uses micro-services, software-defined IoT that is designed to facilitate support for a large network of tenants (multi-tenancy), and load distribution to some more active computational devices than devices on the Internet of Things. This system addresses the problems of things devices such as IoT devices lack virtual and abstract support on restricted and diversified components, there is no distribution of secure software with its hosts and no mechanisms to facilitate it. In software, there is an insufficiency of effective control of access. These resources provide the ability to create a virtual IoT system, and can view one or more components of IoT devices as a restful service. It links IoT device components via digital artifacts and other virtual resources. The virtual resource process can create numerous virtual IoT systems and the existing IoT system by checking its internal state [39].

The virtual resources allow a mechanism for multi-tenancy handling since each tenant has its respective virtual resources, load distribution based on micro-services, access to things in a mechanized way. A permission in the blockchain system is the main challenge edge devices face in enabling secure code deployment. In its infrastructure, Blockchain has a signature or code that edge devices identify to access information from the blockchain without facing any security or safety issues as demonstrated by IBM's ADEPT system. The purpose of the permission-based blockchain approach is to deploy information securely and in this experiment, due to free blockchain service, a lot of fluctuations ranging from 0 to 300 ms are seen. Thus, fluctuating loads result in considerable variation in response time. However, many authors suggest that blockchains based on permission have the potential to store data on virtual resources based on their experimental findings [40].

9.3 Overview of Blockchain & Big data in health care

Big data is a large set of complex data with multiple properties which are analyzed computationally to identify patterns related to it. Big data is a term for a set of data which is so enormous and composite that existing applications for data handling are insufficient. 3Vs (volume, speed, and variety) are used to define the characteristics and dimensions of an enormous set of data or large data. "Volume" refers to size and dimensionality of the data. The data processing speed is marked with "velocity." Finally, "variety" denotes the combination of a number of different data types.

9.3.1 Big Data in Healthcare

Large amounts of data stored together in a specific place known as big data have their own particular properties that computers use to identify the data. Big data characteristics and dimensions are defined with the aid of 3V's (i.e. volume, velocity, and variety). With the advancement in healthcare technology, a large amount of patient data is produced. An electronic medical record as shown in **Figure 9.1**.

Medical records from many sources are available for doctors from many other healthcare devices, such as text, diagnostic images, etc. The two main factors of big data that are present in healthcare today. Big data and data mining are very close to each other in healthcare because drawing an EHR pattern can help doctors to precise disease predictions [41].

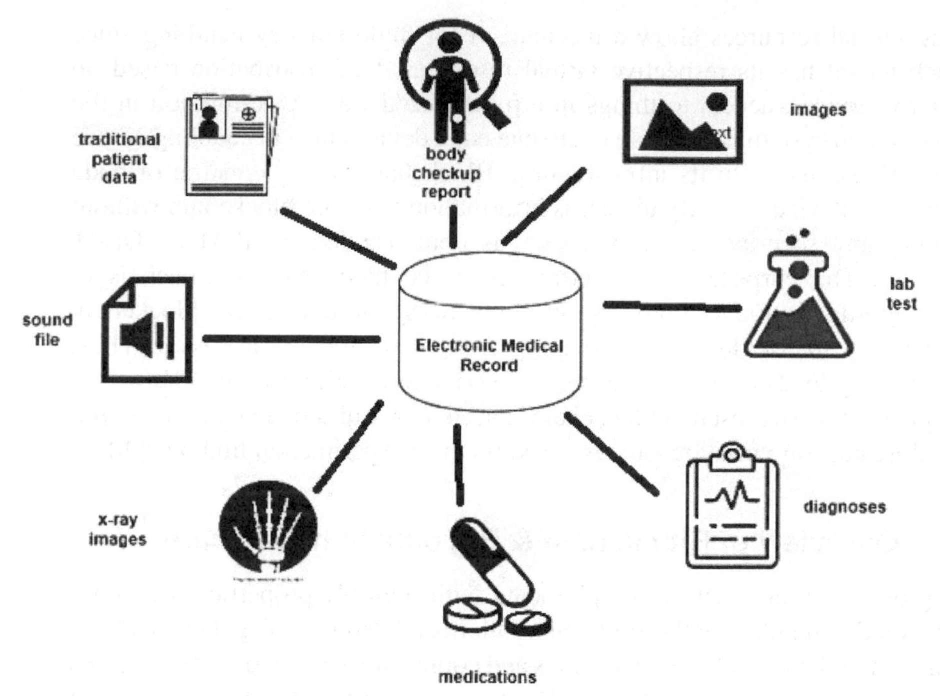

Figure 9.1 Electronic Medical Record

9.3.2 Blockchain in Health Care

Blockchain is an immutable digital data ledger that confidentially displays the transparency of any type of data, as well as data stored in it. It is initially based on the special kind of concept of architecture also known as p2p architecture. Information is stored in various blocks in this technology which have their own unique identity, also known as a hash. This hash is created based on the data that is stored in the block, so every time you do a new hash on the block material. Besides that, each block also contains a hash of the previous section as shown in Figure 9.2. As a result, that block combination is called a blockchain. This interlinking property of blocks gives the data which is one of its best features the special type of security .. Blockchain requires updating the hash values of the whole series to make any changes to a block 's data which makes it impossible to hack and manipulate the secured data.

Blockchain technology has four pillars, on which it depends entirely [42]. Consensus, smart contracts, ledger, and cryptography are the main pillars of these four. Consensus monitors actions within networks and verifies the

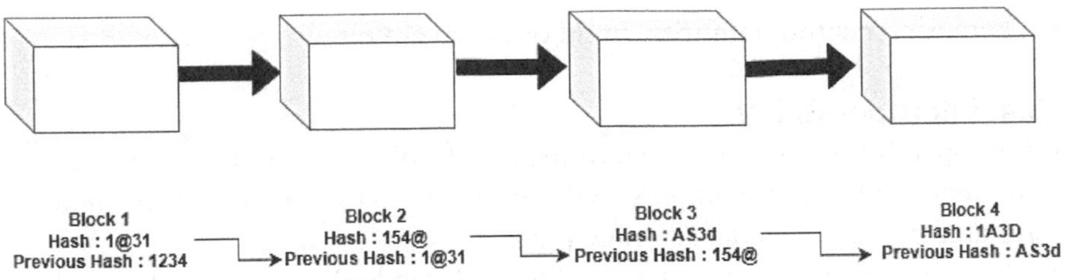

Figure 9.2 Hash formed using blocks

work in them. Smart contracts are used for inspecting and certifying network participants. Ledger is used to provide the entire network transaction structure and cryptography ensures that all the uncorrupted and valid user information in the network and ledger can crack the data. The concept of blockchain in healthcare is very advantageous as it gives the healthcare sector more benefits.

9.3.3 Benefits of Blockchain in Healthcare

The benefits of blockchain in healthcare industry are explained as follows:

9.3.3.1 Master patient indices

- To deal with health care data, sometimes records get duplicated or mismatched [43].
- A scheme of separate electronic health records may be a separate schema for each region coming up with various ways of entering and manipulating the simplest sets of data.
- A blockchain will place the entire data set in an account and not just the primary key.
- The user will search for an address - here may have multiple addresses and multiple keys, but they will all produce to identify one patient.

9.3.3.2 Supply chain management

- Blockchain-based contracts can help health care organizations to monitor supply-demand cycles through their entire life cycle. T
- Transactions take place, whether the contact is successful or there is any delay [44].

9.3.3.3 Claims adjudication

- Blockchain works on a valid- based exchange, claims can be automatically verified where the network agrees in a way to execute the contract.

- There is no central authority, fraud or errors are rare.

9.3.3.4 Interoperability

- Interoperability is the commitment of blockchain that can be implemented using complex API to make EHRs interoperable and reliable processes of data storage [45].
- Blockchain networks are shared safely and standardized with authorized providers.
- Costs and burdens associated with data management are eliminated.

9.3.3.5 Single, longitudinal patient records

- Lab results, longitudinal patient records, diseases register and treatments are achieved through blockchain.
- It includes better ways of providing secure and safe data [46].

9.4 Application of Big Data for Blockchain

Healthcare systems are an architectural paradigm coupled with pervasive sensing and communication technologies that provide the medical environment with multiple benefits. It is an engineered system, where cyber components such as computational hardware and communication network enhance the physical system or process. These components are very tightly integrated, meaning that the functionality of one component depends on the other. Key areas of research are stability, reliability, robustness, security and privacy whilst designing such systems to be smart, efficient and flexible. There are various factors by which we can say that blockchain is applicable for data as discussed below.

9.4.1 Smart Ecosystem

Hospitals, patients, entrepreneurs, along with the data producer, competitors, and suppliers create the healthcare system network. Blockchain technology can create a smart ecosystem for managing health care data. Using P2P (i.e. peer-to-peer) collaboration with blockchain technology can bring about a new era of advanced healthcare economy. Blockchain is a clever way of reducing costs, time and system losses in the management and processing of medical data. Blockchain is a concept of "smart contract" or a smart way of ensuring rules, allowing for any exchange of value [47]. It retains the Contract in a very successful Way.

9.4.2 Digital Trust

From the data analyst to the data provider, patient to physician, everyone wants to have confidence in their business. Blockchain technology can bring this digital trust between big data healthcare dealers. Mattila discussed many factors out of which the need for digital trust between dealers is three main factors, which can only be achieved through a technology, i.e. blockchain identification, traceability, and security. When the data handler wants to share the data with the dealer, a contract exists between the parties to continue to establish the trust [48]. Blockchain technology creates a reliable platform between the digital and physical sectors In contrast, several studies require digital trust in healthcare technology to better analyze and support each patient's data. Jirotka et al. [49] referred to the operating capacity of e-healthcare data and maintained this capacity among Internet stakeholders.

9.4.3 Cybersecurity

Data stored by blockchain technology is immutable. The storage of any transaction, decision, information, and contract is manipulated free technique. The Washington and Pentagon times said the United States military sees this technique as a cybersecurity protective layer. Blockchain technology stores data in distributed ways. By decentralized way of storing information, however, data maintenance can be reduced in this. Nugent et al [50] discussed data distribution and shared that data via an advanced technology called the blockchain. A Maddux study describes the scope of blockchain technology in the Big Data sector in healthcare. Data portability and data distribution is safeguarded. According to Ponemon and IBM, leaking vital healthcare information costs about 380 us dollars per second while the cost of data breaching in the industry sector amounts to only 141 US dollars per instant [51].

9.4.4 Fighting Drugs

Drugs that are both spurious and of very poor quality are present all over the world. By tracking those drugs from where they are produced for end-user feedback, blockchain technology can help maintain medicine standards. According to a report published by the WHO (i.e., the World Health Organization) [52] "one of the ten medical products in low- and medium-income countries is either considered inferior or proved to be wrong." The production of drugs and the distribution chain can be easily tracked by blockchain technology.

9.4.5 Online Accessing of Patient's Data

Blockchain technology provides a double shield, secure platform for managing online medical records. Patients can do this without third party involvement and this is one of the main advantages. The three main capabilities of blockchain technology that make this technology a powerful tool for data management [53]. This technology helps build a network and then creates resources with anonymity. Finally, automatic consent is provided prior to any decision.

9.4.6 Research as well as Development

Massive statistics on health care, health markets are analyzing the genomic data. Blockchain technology can ensure the proper tracking of the quality of information relating to genomics and health [54]. Recently, both the biological and the person's health data are inaccurate and altered which can affect the research quality [55]. If the stakeholders guarantee information excellence, then collaborative research can get more attention.

9.4.7 Management of Data

Big data mining technique has become more complex with the neural networks, deep learning, and artificial intelligence system. Previous data that has been wasted in healthcare can be converted into vital information and further use this vital information for knowledge gaining [56]. The problem behind data handling is privacy and the risk of manipulation. Blockchain guarantees a multilevel data protection system, using decentralization and smart consensus [57]. Information is safe, from health to wearable IoT, in which investors like doctors and so on can be "miners" in this network system. The large data sector in healthcare can use a blockchain to develop data-driven business intelligence [58].

9.4.8 Due to privacy storing of off-chain data

Off-Chain Data gains popularity because of many restrictions on the privacy of customer's data. In general, blockchain has disadvantages due to its decentralized architecture, i.e. it is not able to store data which is very large in quantity. On the contrary, health-care and biological data are naturally large in numbers.

9.4.9 Collaboration of patient data

Patients may set up groups to care for one another. The blockchain technology can maintain the system, as well as monitor it. Patients affected by the associated illnesses feel uncomfortable with the disease.

9.5 Solutions of Blockchain For Big Data in Health Care

Healthcare system opens up plenty of chances to blockchain implementation. Many stakeholders are involved, such as patients, hospitals, doctors, clinical researchers, policy insurers, and suppliers of drugs to name a few. There is a lot of data exchange across the various industry stakeholders. It becomes therefore important that the data is not misused by any entity by illegally sharing the records, manipulating the records, or any other means. There are various challenges in big data in healthcare and the solutions of the blockchain technology depend on various factors. Some of them are discussed in **Table 9.1**.

Table 9.1 Solution of Blockchain For Big Data in Healthcare [59-61].

Factors	Challenges of Big Data in Health Care	Solution of Blockchain
B2B(business to business) Communication	Recently, To make a profit, third parties and middlemen distribute and manipulate data.	To create a boundary for distribution of data in a more secure way, blockchain can remove every third party.
Data fragmentation	In a fragmented decentral way, data is produced. Also separated data is produced by doctors, patients, therapists, and analytics.	Among all the groups (Blockchain Nodes), networks of computers and the connected decentralized to this network can help to create a network.
Data Owner's Privacy	Recently, processors are hidden and owner's identities are open.	technology of Blockchain provides a good way of management of data where identity is open to everyone. Technology of blockchain also helps to save our data from any malicious activity.
Time to time access	Cooperation, time to time access and concurrent analysis.	There is no risk of the manipulation of data in blockchain technology. Moreover, Blockchain can analyses this data.

Continued

Table 9.1 Continued

Factors	Challenges of Big Data in Health Care	Solution of Blockchain
Data Processing Cost	At present, many third parties analyze data with all individuals that increase costs.	To reduce the operation, cost analysis and distribution, with the help of blockchain middlemen less concurrent architecture.
System Scalability	Trust issues (figures and individuals) create unrest in cooperation between parties.	Participation nodes and associated data are verified by blockchain, which eliminates all risks (data authentication and nodes through consensus of algorithm).
Consistency of data	Data handlers should remain closed.	Blockchain can use different varieties of it to process patient data.
Sensor(IoT) Data Handling	To collect and distribute data, thousands of IoT devices are used which are not easy to track as well as handle.	This technology can protect the device of IoT by a special technology known as blockchain.

9.6 Conclusion and Future Scope

From the timeframe of its foundation, Blockchain technology has developed. As part of revolutionary crypto-currency bitcoin technology, the blockchain mechanism has allowed for use in different areas of business and society, such as confidence, transparency, traceability, certification, auditing, etc. We overlooked this technology in the healthcare field very extensively in this chapter. We have also focused on and how to address the use and challenges of these technologies in health care. We also introduced an idea of interoperable and integrated blockchain systems which can function from end to end system Blockchain is an advanced technology and is currently in the innovation stage of its applications in the healthcare industry. As we have already discussed, it is essential that researchers move from the stage of idea to the stage of implementation to make blockchain applications in healthcare systems successful. It is really difficult to measure your actual stage without seeing actual protocols and stopping the functioning of blockchain applications. The interoperability issue in the blockchain is another important and logical issue. The researcher needs to address the possibility of standard protocols with a consensus mechanism and an intelligent contract to facilitate data exchange on several platforms between blockchains. Development of appropriate data storage and processing computational protocols to make the blockchain models scalable in view of other reliabilities. After these challenges have been mitigated, the opportunity would be increased to promote the blockchain in the health system. The proper adoption of blockchain technology will reduce numerous challenges to the existing dimension of health systems and thus result in a healthcare system that is more stable and operational.

References

[1] Chevallereau, B., Carter, G., & Sneha, S. (2019). Voice Biometrics and Blockchain: Secure Interoperable Data Exchange for Healthcare. *Blockchain in Healthcare Today, 2*. doi:10.30953/bhty. v2.119.

[2] Chaudhari, H., Rachuri, R., & Suresh, A. (2020). Trident: Efficient 4PC Framework for Privacy Preserving Machine Learning. *Proceedings 2020 Network and Distributed System Security Symposium*. doi:10.14722/ndss.2020.23005

[3] Mazlan, A. A., Daud, S. M., Sam, S. M., Abas, H., Rasid, S. Z. A., & Yusof, M. F. (2020). Scalability Challenges in Healthcare Blockchain System—A Systematic Review. IEEE Access, 8, 23663–23673. doi: 10.1109/access.2020.2969230

[4] Ahmed, A., Kleiner, M., & Roucoules, L. (2019). Model-Based Interoperability IoT Hub for the Supervision of Smart Gas Distribution Networks. IEEE Systems Journal, 13(2), 1526–1533. doi: 10.1109/jsyst.2018.2851663

[5] Wang, S., Wang, J., Wang, X., Qiu, T., Yuan, Y., Ouyang, L., ... Wang, F.-Y. (2018). Blockchain- Powered Parallel Healthcare Systems Based on the ACP Approach. IEEE Transactions on Computational Social Systems, 5(4), 942–950. doi: 10.1109/tcss.2018.2865526

[6] Li, X., Huang, X., Li, C., Yu, R., & Shu, L. (2019). EdgeCare: Leveraging Edge Computing for Collaborative Data Management in Mobile Healthcare Systems. *IEEE Access, 7*, 22011-22025. doi:10.1109/access.2019.2898265

[7] Wazid, M., Das, A. K., Shetty, S., & Jo, M. (2020). A Tutorial and Future Research for Building a Blockchain-Based Secure Communication Scheme for Internet of Intelligent Things. *IEEE Access, 8*, 88700-88716. doi:10.1109/access.2020.2992467

[8] Esposito C, De Santis A, Tortora G, Chang H, Choo KKR. Blockchain: a panacea for healthcare cloud-based data security and privacy? IEEE Cloud Computing 2018;5(1): 31e7.

[9] Li, P., Xu, C., Jin, H., Hu, C., Luo, Y., Cao, Y., ... Ma, Y. (2020). ChainSDI: A Software-Defined Infrastructure for Regulation-Compliant Home-Based Healthcare Services Secured by Blockchains. *IEEE Systems Journal, 14*(2), 2042-2053. doi:10.1109/jsyst.2019.2937930

[10] D., D. (2020). Survey on Fog Computing: Issues and Challenges. *Journal of Advanced Research in Dynamical and Control Systems, 12*(SP3), 1301–1314. doi: 10.5373/jardcs/v12sp3/20201379

[11] Yang, R., Yu, F. R., Si, P., Yang, Z., & Zhang, Y. (2019). Integrated Blockchain and Edge Computing Systems: A Survey, Some Research Issues and Challenges. *IEEE Communications Surveys & Tutorials, 21*(2), 1508–1532. doi: 10.1109/comst.2019.2894727

[12] Mior, M. J., Salem, K., Aboulnaga, A., & Liu, R. (2017). NoSE: Schema Design for NoSQL Applications. IEEE Transactions on Knowledge and Data Engineering, 29(10), 2275–2289. doi: 10.1109/tkde.2017.2722412

[13] Misaki, M., Tsuda, T., Inoue, S., Sato, S., Kayahara, A., & Imai, S.-I. (2017). Distributed Database and Application Architecture for Big Data Solutions. IEEE Transactions on Semiconductor Manufacturing, 30(4), 328– 332. doi: 10.1109/tsm.2017.2750183

[14] Dayalan, M. (2018). MapReduce: Simplified Data Processing on Large Cluster. International Journal of Research and Engineering, 5(5), 399–403. doi: 10.21276/ijre.2018.5.5.4

[15] Ismail, L., Materwala, H., & Zeadally, S. (2019). Lightweight Blockchain for Healthcare. *IEEE Access, 7*, 149935-149951. doi:10.1109/access.2019.2947613

[16] Mazlan, A. A., Daud, S. M., Sam, S. M., Abas, H., Rasid, S. Z., & Yusof, M. F. (2020). Scalability Challenges in Healthcare Blockchain System—A Systematic Review. *IEEE Access, 8*, 23663-23673. doi:10.1109/access.2020.2969230

[17] Chukwu, E., & Garg, L. (2020). A Systematic Review of Blockchain in Healthcare: Frameworks, Prototypes, and Implementations. *IEEE Access*, *8*, 21196–21214. doi: 10.1109/access.2020.2969881

[18] Rajnish, D. R. (2019). Securing Healthcare Records using Blockchain Technology. *SSRN Electronic Journal*. doi: 10.2139/ssrn.3349590

[19] Rahman, M. A., Hossain, M. S., Loukas, G., Hassanain, E., Rahman, S. S., Alhamid, M. F., & Guizani, M. (2018). Blockchain-Based Mobile Edge Computing Framework for Secure Therapy Applications. *IEEE Access*, *6*, 72469–72478. doi: 10.1109/access.2018.2881246

[20] Chukwu, E., & Garg, L. (2020). A Systematic Review of Blockchain in Healthcare: Frameworks, Prototypes, and Implementations. *IEEE Access, 8*, 21196-21214. doi:10.1109/access.2020.2969881

[21] Syed, T. A., Alzahrani, A., Jan, S., Siddiqui, M. S., Nadeem, A., & Alghamdi, T. (2019). A Comparative Analysis of Blockchain Architecture and its Applications: Problems and Recommendations. *IEEE Access, 7*, 176838-176869. doi:10.1109/access.2019.2957660

[22] Hasan, H. R., Salah, K., Jayaraman, R., Omar, M., Yaqoob, I., Pesic, S., ... Boscovic, D. (2020). A Blockchain-Based Approach for the Creation of Digital Twins. *IEEE Access*, *8*, 34113–34126. doi: 10.1109/access.2020.2974810

[23] Fernandez-Carames, T. M., & Fraga-Lamas, P. (2020). Towards Post-Quantum Blockchain: A Review on Blockchain Cryptography Resistant to Quantum Computing Attacks. *IEEE Access*, *8*, 21091–21116. doi: 10.1109/access.2020.2968985

[24] Sharma, S., Ghanshala, K. K., & Mohan, S. (2019). Blockchain-Based Internet of Vehicles (IoV): An Efficient Secure Ad Hoc Vehicular Networking Architecture. *2019 IEEE 2nd 5G World Forum (5GWF)*. doi: 10.1109/5gwf.2019.8911664

[25] Kumar, A., Krishnamurthi, R., Nayyar, A., Sharma, K., Grover, V., & Hossain, E. (2020). A Novel Smart Healthcare Design, Simulation, and Implementation Using Healthcare 4.0 Processes. *IEEE Access, 8*, 118433-118471. doi:10.1109/access.2020.3004790

[26] Saini, H., Bhushan, B., Arora, A., & Kaur, A. (2019). Security vulnerabilities in Information communication technology: Blockchain to the rescue (A survey on Blockchain Technology). 2019 2nd International Conference on Intelligent Computing, Instrumentation and Control Technologies (ICICICT). DOI: 10.1109/icicict46008.2019.8993229

[27] Chen, W., Chen, Y., Chen, X., & Zheng, Z. (2019). Toward Secure Data Sharing for the IoV: A Quality- Driven Incentive Mechanism with On-Chain and Off-Chain Guarantees. *IEEE Internet of Things Journal*, 1–1. doi: 10.1109/jiot.2019.2946611

[28] Wang, Y., Zhang, A., Zhang, P., & Wang, H. (2019). Cloud-Assisted EHR Sharing With Security and Privacy Preservation via Consortium Blockchain. *IEEE Access*, *7*, 136704–136719. doi: 10.1109/access.2019.2943153

[29] Li, H., & Han, D. (2019). EduRSS: A Blockchain-Based Educational Records Secure Storage and Sharing Scheme. *IEEE Access*, *7*, 179273–179289. doi: 10.1109/access.2019.2956157

[30] https://www.media.mit.edu/publications/medrec-whitepaper/

[31] R. Guo, H. Shi, Q. Zhao, D. Zheng, Secure attribute-based signature scheme with multiple authorities for blockchain in electronic health records systems, IEEE Access 6 (2018) 11676–11686.

[32] Bhushan, B., Khamparia, A., Sagayam, K. M., Sharma, S. K., Ahad, M. A., & Debnath, N. C. (2020). Blockchain for smart cities: A review of architectures, integration trends and future research directions. Sustainable Cities and Society, 61, 102360. DOI: 10.1016/j.scs.2020.102360

[33] Ghadamyari, M., & Samet, S. (2020). Decentralized Electronic Health Records (DEHR): A Privacy-preserving Consortium Blockchain Model for Managing Electronic Health Records. *Proceedings of the 6th International Conference on Information and Communication Technologies for Ageing Well and E-Health.* doi:10.5220/0009398101990204

[34] A, K., & R, C. (2020). An Efficient Authentication Scheme For Block Chain-Based Electronic Health Records. *International Journal of Engineering Applied Sciences and Technology*, *04*(09), 465–470. doi: 10.33564/ijeast.2020.v04i09.063

[35] Sharma, T., Satija, S., & Bhushan, B. (2019). Unifying Blockchian and IoT: Security Requirements, Challenges, Applications and Future Trends. 2019 International Conference on Computing, Communication, and Intelligent Systems (ICCCIS). DOI: 10.1109/icccis48478.2019.8974552

[36] Lv, P., Wang, L., Zhu, H., Deng, W., & Gu, L. (2019). An IOT-Oriented Privacy-Preserving Publish/Subscribe Model Over Blockchains. *IEEE Access*, *7*, 41309–41314. doi: 10.1109/access.2019.2907599

[37] Belotti, M., Bozic, N., Pujolle, G., & Secci, S. (2019). A Vademecum on Blockchain Technologies: When, Which, and How. IEEE Communications Surveys & Tutorials, 21(4), 3796–3838. doi: 10.1109/comst.2019.2928178

[38] Varshney, T., Sharma, N., Kaushik, I., & Bhushan, B. (2019). Authentication & Encryption Based Security Services in Blockchain Technology. 2019 International Conference on Computing, Communication, and Intelligent Systems (ICCCIS). DOI: 10.1109/icccis48478.2019.8974500

[39] Soni, S., & Bhushan, B. (2019). A Comprehensive survey on Blockchain: Working, security analysis, privacy threats and potential applications. 2019 2nd International Conference on Intelligent Computing, Instrumentation and Control Technologies (ICICICT). DOI: 10.1109/icicict46008.2019.8993210

[40] Mencias, A. N., Dillenberger, D., Novotny, P., Toth, F., Morris, T. E., Paprotski, V., ... Carbarnes, E. (2018). An optimized blockchain solution for the IBM z14. *IBM Journal of Research and Development, 62*(2/3). doi: 10.1147/jrd.2018.2795889

[41] Viswanathan, R., D. Dasgupta, and S. R. Govindaswamy. "Blockchain Solution Reference Architecture (BSRA)." *IBM Journal of Research and Development* 63, no. 2/3 (2019). https://doi.org/10.1147/jrd.2019.2913629.

[42] Sinha, P., Rai, A. K., & Bhushan, B. (2019). Information Security threats and attacks with conceivable counteraction. 2019 2nd International Conference on Intelligent Computing, Instrumentation and Control Technologies (ICICICT). DOI: 10.1109/icicict46008.2019.8993384

[43] Zhang, Xiaohong, and Xiaofeng Chen. "Data Security Sharing and Storage Based on a Consortium Blockchain in a Vehicular Ad-Hoc Network." *IEEE Access* 7 (2019): 58241–54. https://doi.org/10.1109/access.2018.2890736.

[44] Gupta, S., Sinha, S., & Bhushan, B. (2020). Emergence of Blockchain Technology: Fundamentals, Working and its Various Implementations. SSRN Electronic Journal. DOI:10.2139/ssrn.35695

[45] Singh, Priya, Pooja Khanna, and Sachin Kumar. "Communication Architecture for Vehicular Ad Hoc Networks, with Blockchain Security." *2020 International Conference on Computation, Automation and Knowledge Management (ICCAKM)*, 2020. https://doi.org/10.1109/iccakm46823.2020.9051499.

[46] Arora, D., Gautham, S., Gupta, H., & Bhushan, B. (2019). Blockchain-based Security Solutions to Preserve Data Privacy and Integrity. 2019 International Conference on Computing, Communication, and Intelligent Systems (ICCCIS). DOI: 10.1109/icccis48478.2019.8974503.

[47] Feng, M., Zheng, J., Ren, J., Hussain, A., Li, X., Xi, Y., & Liu, Q. (2019). Big Data Analytics and Mining for Effective Visualization and Trends Forecasting of Crime Data. IEEE Access, 7, 106111–106123. doi: 10.1109/access.2019.2930410

[48] Varshney, T., Sharma, N., Kaushik, I., & Bhushan, B. (2019). Architectural Model of Security Threats& their Countermeasures in IoT. 2019 International Conference on Computing, Communication, and Intelligent Systems (ICCCIS). DOI:10.1109/icccis48478.2019.8974544

[49] Jirotka, M., Procter, R., Hartswood, M., Slack, R., Simpson, A., Coopmans, C., ... Voss, A. (2005). Collaboration and Trust in Healthcare Innovation: The eDiaMoND Case Study. Computer Supported Cooperative Work (CSCW), 14(4), 369-398. doi:10.1007/s10606-005-9001-0.

[50] Nugent, T., Upton, D., & Cimpoesu, M. (2016). Improving data transparency in clinical trials using blockchain smart contracts. F1000Research, 5, 2541. doi:10.12688/f1000research.9756.1.

[51] NIST Big Data Interoperability Framework: volume 1, definitions, version 2. (2018). doi: 10.6028/nist.sp.1500- 1r1.

[52] Houtan, B., Hafid, A. S., & Makrakis, D. (2020). A Survey on Blockchain-Based Self-Sovereign Patient Identity in Healthcare. *IEEE Access, 8*, 90478-90494. doi:10.1109/access.2020.2994090

[53] Kashyap, R. (2019). Big Data Analytics Challenges and Solutions. Big Data Analytics for Intelligent Healthcare Management, 19–41. doi: 10.1016/b978-0-12-818146-1.00002-7

[54] Tiwari, A., Sharma, N., Kaushik, I., & Tiwari, R. (2019). Privacy Issues & Security Techniques in Big Data. 2019 International Conference on Computing, Communication, and Intelligent Systems (ICCCIS). doi: 10.1109/icccis48478.2019.8974511

[55] Zghaibeh, M., Farooq, U., Hasan, N. U., & Baig, I. (2020). SHealth: A Blockchain-Based Health System With Smart Contracts Capabilities. IEEE Access, 8, 70030–70043. doi: 10.1109/access.2020.2986789

[56] Bikakis, N. (2019). Big Data Visualization Tools. Encyclopedia of Big Data Technologies, 336–340. doi: 10.1007/978-3-319-77525-8_109

[57] Tiwari, A., Sharma, N., Kaushik, I., & Tiwari, R. (2019, October). Privacy Issues & Security Techniques in Big Data. In 2019 International Conference on Computing, Communication, and Intelligent Systems (ICCCIS) (pp. 51-56). IEEE.

[58] Parmar, N., Refai, H., & Runolfsson, T. (2020). A Survey on the Methods and Results of Data-Driven Koopman Analysis in the Visualization of Dynamical Systems. IEEE Transactions on Big Data, 1–1. doi: 10.1109/tbdata.2020.2980849.

[59] Goyal, S., Sharma, N., Bhushan, B., Shankar, A., & Sagayam, M. IoT Enabled Technology in Secured Healthcare: Applications, Challenges and Future Directions. In *Cognitive Internet of Medical Things for Smart Healthcare* (pp. 25-48). Springer, Cham.

[60] Kumar, S., & Singh, M. (2019). A novel clustering technique for efficient clustering of Big Data in Hadoop Ecosystem. Big Data Mining and Analytics, 2(4), 240–247. doi: 10.26599/bdma.2018.9020037

[61] Zhong, J., Han, T., Lotfi, A., Cangelosi, A., & Liu, X. (2019). Bridging the Gap between Robotic Applications and Computational Intelligence in Domestic Robotics. 2019 IEEE Symposium Series on Computational Intelligence (SSCI). doi: 10.1109/ssci44817.2019.9002883

10

Cloud Resource Management for Network Cameras

Hemanta Kumar Bhuyan[1*] **and Subhendu Kumar Pani**[2]

[1]Department of Information Technology, Vignan's Foundation for Science, Technology & Research (Deemed to be University), Guntur, Andhra Pradesh, India
[2]Principal, Krupajal Computer Academy, India
E-mail: hmb.bhuyan@gmail.com, skpani.india@gmail.com
*Corresponding Author

Abstract

Most of the network cameras flow dynamic and adequate visual data from several locations such as traffic, check gate/tollgate, natural scenes, etc.These data from high resolution network cameras can be used in several applications that help to find the needed significant resources. The cloud vendors provide the required resources as per the hourly cost which is not always so affordable for several applications using cameras and instances. This problem can be solved using bin packing and heuristic algorithm. Thus, the allocation and deallocation of instances are maintained by cloud resource manager, by which the cost is controlled accordingly. The performance of this framework is considered based on multiple analyses of programs, moving object detection, feature tracking, and large number of images.

Keywords: Cloud resource management, network camera, dynamic bin packing (DBP), resource monitoring.

10.1 Introduction

Over the last few decades, the large number of sensitive network cameras (NetCams) has been issued to catch different images and video frames around the globe. The million number of NetCams are grown and produce huge amount of terabyte (TB) per day [1,2]. In addition,different cameras maintain individual frame rates and resolution as their capacity. These cameras have been used to stream the real-time visual data from public sources or a variety of environments. Thus, these image or video data always help to diversify uses such as super vision of public networking, surveillance, weather detection, etc.

For high resolution and high speed of frame rates of image or video, high configuration of electro-computing device is needed to handle all kinds of visual data based on considerable resources. Cloud computing provides such kinds of facilities for most of the resources. Thus, different cloud service providers or vendors suggest implementing diversity application with different needed resources like high configuring computing network system to build sensitive cameras. But for the implementation of such resources, financial support is needed to pay in a pay-as-you-go manner as per cloud vendor's agreement/rule, i.e., based on use of application, customers choose payment option as above.

The cost of using several instances through cloud is a challenging problem. In each day, the real-time visual data streams are needed to analyze for several incidents from different places. It is possible only through cloud resources. Cloud resources are managed for lessening the price rate for visual data flows during the huge number of cameras are working parallel to perform well. To simplify the above difficulties, two categories of solutions are considered: (a) reserved allocation and (b) extended source.

(a) Reserved allocation:
This part is important for allocating the resources as per cloud instances. The manager can take the decision based on the following manner:

(i) use the selective cloud instances;
(ii) allocate number of instances;
(iii) assign instances which cameras.

(b) Extended source:
In this category, it considers utilization as well as the scale of resources. As per category (a), similar manner is decided by the manager as follows:

(i) resources handle through diversity of utilization;
(ii) extended number of operational instances.

The above choices are influenced to process visual stream using cloud environment through number of issues as follows:

(a) program descriptions;
(b) needed visual data ratios;
(c) quality image or videos through cameras;
(d) visual details from NetCam;
(e) price rates of utilization.

Thus, the resource manager may pay attention to provide the facilities of the above factors before releasing. But the resource manager always tries to lessen the overall price rate of each data flow through NetCam based on users' profit with well concert. Generally, the total price rate is calculated as per hourly use of any applications. But the resource manager always decides on the basis of the following factors for the possibility of providing facilities. So, it:

(i) calculates high frame rates and resolutions approximately on needed source for data flows through NetCam;
(ii) plans the resource allocation problem by bin packing problem and find the solutions by means of a heuristic algorithm;
(iii) regularly supervises the instances through cloud for controlling utilization;
(iv) updates allocation of instances as needed;
(v) transfers data flows to get good benefits with lessened price rates.

Thus, most of organization uses CAM2for individual or official work to interact with global customers. This camera is helpful to all researchers for better utilization of study. Since it is a network-based camera, it needs cloud environment to use in diversity purpose. The computational design in cloud may extend to fulfill the several program analyses. Therefore, it considers CAM2 which is integration with regular utilization of several cameras as per needed features of image and videos. CAM2 is used for image and video data flows through cloud environment.

Many open websites are used for public [3] for sharing camera data over cloud. The performance on these testing aspects is considered based on experiments that evaluate the consequence factors such as resolutions and kind of instances as per choice of resource management. Different parts and resources are explained in the sub sequent section.

10.2 Resource Analysis

10.2.1 Network Cameras

Definition: NetCam is also called an Internet Protocol camera or IP camera. It is a special kind of camera for capturing videos and images by accepting digital data and forwards data through network.

Generally, this camera is used for surveillance and processing all kinds of images and videos accordingly. It does not need a local recording device like analog closed-circuit television (CCTV) cameras, but a local area network is needed. All kinds of images and videos can directly entry over cloud through NetCam. The basic design of NetCam maintains a central network video recorder(NVR) to manage several capturing video alarms. Sometimes, this kind of working operation is considered without NVR and with capable remote storage media. Different vendors are developed as per cost and suitability for customers.

Different sources are available to collect several images or video details through NetCams. For example, Archive of Many Outdoor Scenes (AMOS) [4] is a huge data base that holds over eight billion images collected through more than thirty thousand cameras ever since 2006. Many visual data are also available in several websites for public.

Readers can get many applications on NetCams such as surveillance [5] and weather detection [6]. Many systems have been built to analyze the visual data from NetCams. The structure of spatio-temporal analysis and tracking system through camera network is elaborated in [7] and[8]. The system is designed in such a way that it supplies several images and videos through cameras based on cloud environment with dynamic applications.

10.2.2 Resource Management on Cloud Environment

Cloud computing is always engaged through several applicable computation with stock-up sources as per servicing cost. The multitenancy is another feature of cloud computing to activate several resources and price rate distribution along with huge number of customers. Based on huge data set analysis, it practices data management based on good performance by automatic extendable way in cloud computing. Several structural works come out to exercise formation of data updating through a number of cloud resources. Therefore, it always manages different resources as per customer demand and facilities given by service providers. When demand and facilities donot match, the service providers or vendors try to fulfill the corresponding demands to extend the structure workflows through several aspects such as performance, storage, consistency, and accessibility through several deployed cloud computation.

Many applications are utilized through several resources using cloud environment in different direction as customer requirements efficiently. There,source management is designed to operate the optimal resources through data centers [9]. Several resource methods are used to solve several stipulated customer facility problems through either physical or virtual machine (VM) based resource utilization.

The role of resource management is very significant in cloud computing system to solve several problems, when it faces diverse vital situations such as extensions of dynamic data centers, dissimilar kinds of resources, inter- or intra-dependency among the resources, dynamic loads, and customers' demand to maintain cloud ecosystem. In addition, resource management tries to cover several possible facilities with corresponding resources with assigned tasks. The approaches of resource management system in cloud are shown in Figure 10.1.

The resource management maintains three approaches in cloud computing system as follows:

(a) Provisioned resources: These resources are used for the needed assigned task recognized by customers through quality of service (QoS). In these facilities, it identifies the needed resources and implements the workload. The QoS assigns suitable resources for task as per needed application.In this case, customers cooperate and access the necessary QoS of task through cloud portal after validation. The resource information center(RIC) maintains details of all resources with assigned workload as mentioned by user. The resource provisioning agent (RPA) verifies existing resources for providing the facilities to users. After verifying all requests, validation, and availability of resources, the workloads are forwarded to the resource scheduler. Finally, the workload resource manager (WRM) forwards the correct information on resources to the RPA and also the same to the cloud user.

(b) Scheduling resources: This part consists of drawing, allocating, and implementing several workloads using different resources chosen from provisioned resources phase. This phase maps with choosing resources as per QoS requirements and also use service level agreement (SLA) with the condition of lessening cost and computational time of each

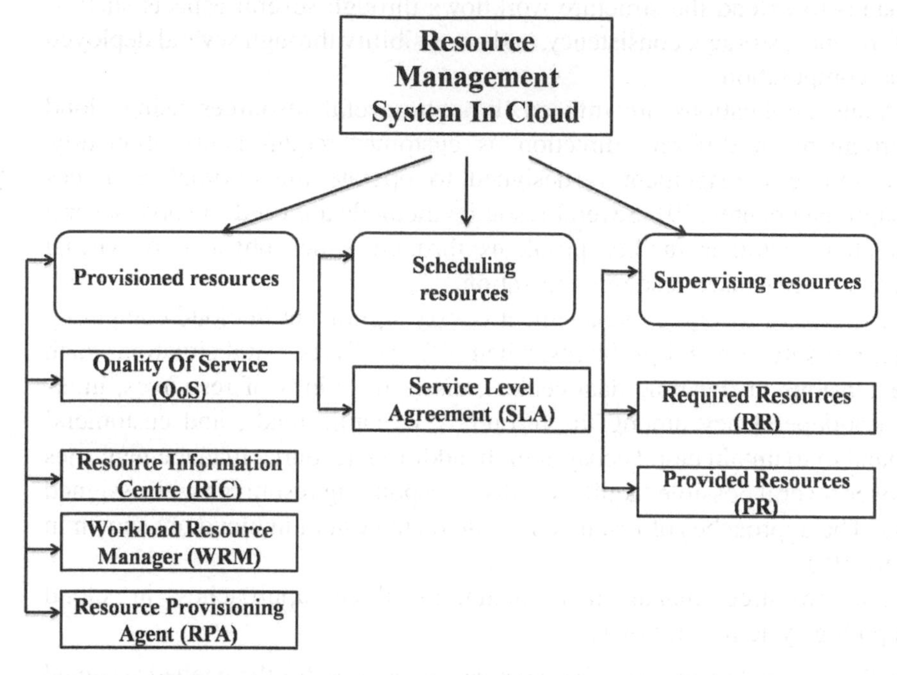

Figure 10.1 Three approaches of resource management system in cloud.

instance. The detection and selection of resources is emphasized by this part through SLA.

(c) Supervising resources:This part is a balancing phase to obtain better performance on assigned resources. The role of SLA is important to supervise frequently based on agreement between customers and providers to avoid any deviations of services, quality of applications, etc. It tries to lessen the deviation of several services and qualities. Further, resource supervising looks after the role QoS needs such as security, availability, and performance This part is always verified and measurements the difference between quantity of required resources (RR) and provided resources (PR). As per scheduling resources, resources are allotted as per availability of resources,by which it controls all kinds of resource-related activities.

The role of network computing is very important to update the scalability and lessen the delay of data [10], [11]. This computational practice does not use individuals due to:(a) lack of control by individual for numerous distributed cameras with local networks throughout the globeand (b) the high cost for individual.Therefore, it is very carefully used for several computations. Some authors like Sharma *et al.* [12] tried to make a special approach for lessening the design cost using optimization technique for any cloud applications. The system estimates various kinds of instances as per onlythe total number of requests, not individuals.The system utilizes a dispatcher to control all loads for web applications.

It focuses to lessen numerous physical machines and develops a set of VMs for video surveillance services [13]. Based on simulations, the cost varies as per applications or uses. Thus, the overall cost is lessened on cloud instances for diverse applications. The solution is developed using vector bin packing formulation for various instances and generates cost charge accordingly. It used a heuristic resource provisioning algorithm for data flowing in the cloud [14]. They assume that:

(i) They are CPU-based applications.
(ii) The execution time may be updated as per CPU cycles for certain VMs.
(iii) The infrastructure cost for the VM is determined based on allocated CPU cycles.

A few authors developed the approaches to lessen the cost for image and video streams from NetCam[15] [16] [17]. Itused the methods for resource allocation problem as a 1D variable sized bin packing problem [17].It means to develop higher resolution cameras based on several resource allocations.It also helps to analyze allocation procedure for managing frame rates with several evaluations on camera resolutions, visual details, and transferring frame rates as per resource requirements.

10.2.3 Image and Video Analysis

Nowadays, digital data is a very significant data for all users as their diverse applications. For safety or research point of view, most of the customers use NetCam for their own requirements.

- To see images or videos, anybody can easily get the right information and use it accordingly.
- It generates quality of information through the digital medium.
- It compares the actual happening and what is being shown through other media.
- It can use various purposes as user's requirements.

- Images and videos can be collected from several sources and distributed among known people.

Sometimes, the above concept is not so directly useful because advanced technology is used for NetCam to capture a variety of images and videos. Several images and videos are generated by NetCam with details which are useful for analysis as user requirements. The network communication maintains properly to manage much information of both images and videos when those are distributed over the Internet. It needs several techniques to transfer visual data with security, by which nobody disturbs the distributed visual data. Thus, it considered image and video analysis based on suitable approaches that include:

- video sequential break down of recording pictures;
- video perception discovery;
- video activities finding with numbering;
- visual recording quality with assessment;
- visual application tools for organizing and synchronizing image and video collections, clustering, distribution, and selection.

Further, machine learning approaches hold upseveral analyses that include:

- deep learning techniques for visual data capturing;
- visual learning on vagueness data;
- learning for classification approaches;
- feature extraction of visual data.

The resource manager utilized visual analysis approaches for background subtraction[24] [25], feature detection [26], feature tracking [27],and human detection [28]. It uses the special approaches through OpenCV [29] execution. Moreover, the reader can access the important concept that is available in [56].

10.3 Cloud Resource Management Problems

The video streams and video frames are an important part of NetCams through the Internet. To activate those parts, cloud resource is needed to manage on the Internet. It can consider the following example for managing cloud resources:different actions are considered for drivers when they travel from one place to another in a city during weather conditions. They execute the following analysis programs during that time:

(i) The frame rates are used for vehicle movement through vehicle tracking program where a vehicle can easily be tracked by frames.

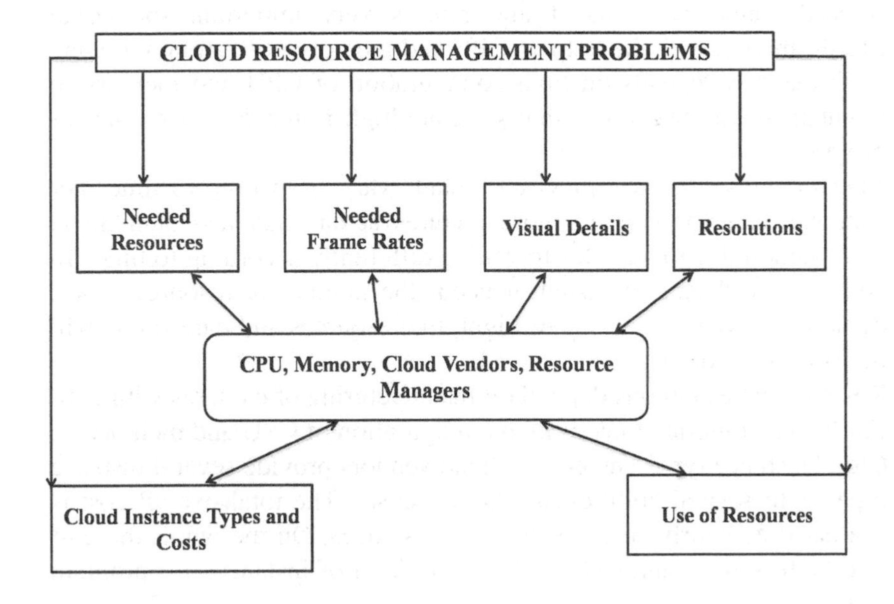

Figure 10.2 Cloud resource management with factors.

(ii) Similarly, the weather can be detected through image streams with frame rates.

Generally, this kind of program does not need high frame rate due to unchanged weather conditions with repeated frames. Several executions can be considered, but it will be very cost effective due to changes of weather and vehicle movement regularly. The analyzer decides the possible reducing cost on frame rates using cloud. The cloud resource management problems are described in Figure 10.2.

The resource manager can take many challenges to solve these kinds of problems. It can describe several aspects on this kind of problem analysis, such as: needed resources, frame rates, camera resolutions, details of visual, cloud instances, costs etc. Thus, the resource manager can analyze and take right decisions using the following factors to solve visual frame rates on cloud environment problems:

(1) Needed resources: the resources are needed for several corresponding related activities. Most of the resource analyses used are CPU intensive as well as memory intensive with distributed manner.

(2) Needed frame rates: the frame rate is very important for visual transformation with the help of CPU and memory utilization. Growing of frame rate depends on high configuration of CPU and memory to maintain stable data flows that generate high frame rates for analysis program.
(3) Visual details: the camera visual details vary from time to time. For example, cameras capture more pictures at day and less at night at any location due to vehicles travelling differently according to time. To describe the details of visual, it needs the number of resources.Thus, tracking the vehicles at day or night, the resources and data frame will be varied over time.
(4) Resolutions:resolutions depend on manufacturing of cameras with cost. For high resolution, it needs high configuration of CPU and memory.
(5) Cloud instance types and costs: Cloud vendors provide several instance types with several abilities and hourly costs. The total overall cost is evaluated as hourly costs of the used instances. On the whole,the cost will be lessened cautiously with correct kind of instances for program analyses.
(6) Use of resources: the recent instances may be reutilized for running fresh program analysis as per the needed resource utilization of corresponding instances. Updating the executing instances and transferring data flows among instances can be considered within received levels.

10.4 Cloud Resource Manager

The role of the cloud resource manager is important to manage the resources properly; otherwise, cost will be varied for customers to access the cloud resources. It decides to handle resource allocation and extension of cloud resources to lessen the cost for benefits of the customer through numerous NetCams concurrently. The resources manager focuses to reach the demanding performance based on the following stages of resource allocation practices:

(1) In the first stage, calculate approximately the needed resources for the data flow from each camera. This part is well calculated for current and future practices for equivalent program.
(2) Based on cloud instances allocation, the whole cost will be reduced accordingly.

The whole cost is always calculated based on hourly costs of the used instances.Theresource manager always supervises and updates the number of operating instances.

10.4.1 Evaluation of Performance

The resource manager aims to perform several analysis programs through data flows with frame rates from numerous cameras.The performance is evaluated as the ratio between the genuine frame rate and the needed frame rate. The performance η_c is formulated for single camera (c) as

$$\eta_c = f_c/f$$

where f_c is the genuine frame rate and f is the needed frame rate of camera c. Thus, for all the cameras, the on-the-whole performance is denoted by η (ϵ[0, 1]) and can be evaluated as

$$\eta = \frac{1}{N} \sum_{c=1}^{N} \frac{f_c}{f}$$

where N is the number of cameras.

This performance is influenced by several factors. Sometimes, few factors are restricted by the resource manager, such as unpredicted network circumstances and unauthenticated user's camera data. Thus,the resource manager controls all resource utilization.For over utilization of resources, the performance lessens accordingly. There source manager tries to continue the performance to reach 100% as per availability of resources.

10.4.2 View of Resource Requirements

Initially, resource manager have no idea to develop a wide range of programs. The manager runs each program based on the needed frame rate through data flows collected from multiple cameras. It supervises the CPU and memory utilization for the duration of the analysis. This helps the manager to calculate approximately the needed resources for analyzing a single data stream.

The numerous data streams are used for experimentally executing where the experiment is used for prospect executions of equivalent analysis program. It maintains the good relationship CPU and memory for each frame rate.

Camera resolutions depend on important manufacturing material. For strongly analyzing data streams, it needs high resolution cameras with needed resources.The time complexity and the space complexity also help to generate good quality of resolutions.

Cloud instance types maintain several abilities and hourly costs. The hourly cost of an instance type and its capabilities are proportional to each other. The cost is evaluated as per configuration needed for system. Resources used in [15], [16],and [30] show that smaller instances based on less CPU cores and less memory are more cost-effective than larger ones. As per this inspection, the manager chooses smaller instance types. One example is considered as to how the manager utilizes a single instance based on the following practices:

1. Choose equal resolution of number of cameras for execution of the program.
2. Execute all programs based on needed frame rate on the data streams collected from multiple cameras.
3. Supervise the CPU and memory utilization.
4. Calculate the needed resources to analyze the data stream from a single camera.
5. Generate equal resolution,if different cameras contain different resolutions.

10.5 Bin Packing

In the bin packing problem, several volumes of items are packed into a certain number of bins or containers. The decision problem on a precise number of bins is NP-complete. Several problems are involved with bin packing problems, such as 2D packing, linear packing, packing by cost, and so on. Several examples can be considered such as filling up containers, loading trucks with weight capacity constraints, creating file backups in media, etc.

The best example can be considered as knapsack problem. In this problem, the items are recognized by its volume and its price. Both volume and its price are managed through containers. This problem generates maximum values as per items in the container.

In cloud computing system, its use is identified by VMs for occupying space. The techniques of space adjustment for items are very important. This variant is known as VM packing. It is very significant in managing space through VMs in a server. It manages whole memory when pages are shared by the VMs for storing. For random sharing space,the bin packing problem is difficult to be even approximately solved.

In the standard bin packing problem, it focuses on several sizes of objects lessen into unit size of bins to avoid size of bin packing difficulties [18]. This type of problem is NP-hard [19]. Several observations are considered as:

(i) Variable sized bin packing [20] permits several sizes of bins and tries lessening the entire size of bins for using.
(ii) Vector bin packing [21], [22] utilizes more than one dimension vector with size of each object or bin. In any bin, the total size of the objects will never be exceeded than anyone as per dimension. Vector bin packing with heterogeneous bins [23] permits bins for several sizes and costs. Thus, it focuses to lessen the overall cost of the used bins.

The integer linear programming techniques based on several variable sizes of bin packing generates2D vector bin packing problem using dissimilar bins with the following analyses:

(i) Bin packing problems are used for job assignment and resource allocation.
(ii) 2D bin packing is always utilized to control resources such as CPU and memory.
(iii) The resource manager permits heterogeneity of the bins for several cloud instances based on various abilities with costs.
(iv) It needs more observation on traditional solutions for bin packing problems.

Han *et al.* [23] used heuristic algorithm to solve the 2D vector bin packing problem based on its performance. This algorithm is much faster than other algorithms [23].Other algorithms can be merged to the resource manager exclusive of disturbing its design, i.e., performance.

10.5.1 Analysis of Dynamic Bin Packing

Dynamic bin packing (DBP) is an alternative of classical bin packing where items come in and go out arbitrarily. The traditional DBP tries to lessen the highest number of bins utilized in the packing. The new type of DBP known as MinTotal DBP always tries to lessen the entire cost of bins as per time. It also used the dispatching problem based on cloud gaming systems. It describes several packing algorithms such as First Fit, Best Fit, and Any Fit packing algorithms for the MinTotal DBP problem. The above algorithms maintain the competitive ratio for any Fit packing less than $\mu + 1$, where μ is defined as the ratio between maximum to minimum item duration. The μ is not so effective for Best Fit packing.

The combinatorial optimization problem for bin packing is described in [40] and[42]. In this type of classical problem, it focuses the minimum number of bins for packed items where the total size of the items in each bin does not exceed the bin capacity. The DBP maintains arrival time and departure time for all items [41] and also use resource consolidation problems in cloud computing [48, 54, 57].

The MinTotal DBP problem is basically focused on cloud gaming systems, where computer games execute on controlling cloud servers. All players act together with the games through networked thin clients [46]. The cloud servers execute game instances such as render the 3D graphics, encode them into 2D videos, and stream them to the clients. The clients decode those and exhibit the video streams and scalethecloud computing infrastructure and deal with several researchers [51]. The cloud gaming is provided by different cloud vendors such as GaiKai [32], OnLive [33], and Stream MyGame [34], and also benefits the companies [31].

The approach is also adopted by different service providers like GaiKai and OnLive [55]. Generally,public clouds are used in cloud gaming systems using game servers with VM for minimizing the cost. The online MinTotal DBP problem also helps to analyze this kind of gaming problem.

10.5.2 MinTotal DBP Problem

Cloud gaming systems have been put into practice for both commercial exercise as well as research applications [34], [46], [33]. In addition, several active works have emphasized to determine the performance of cloud gaming systems [39], [53]. But the resource management does not focus on cloud gaming concerns. Figure 10.3 is described as follows based on several bin packing problems. The MinTotal DBP problems are involved in the following research topics:

(a) classical bin packing problem;
(b) DBP problem;
(c) interval scheduling problem.

(a) Classical bin packing problem: It focuses to generate a set of items based on less number of bins. The problem and its dissimilarities are analyzed through both the offline and online versions [40], [44]. The offline version of this problem is known as NP-hard [45]. For online version, all items are assigned to a bin without permission of movement from one to another bin. Thus, the upper and lower bound can be detected

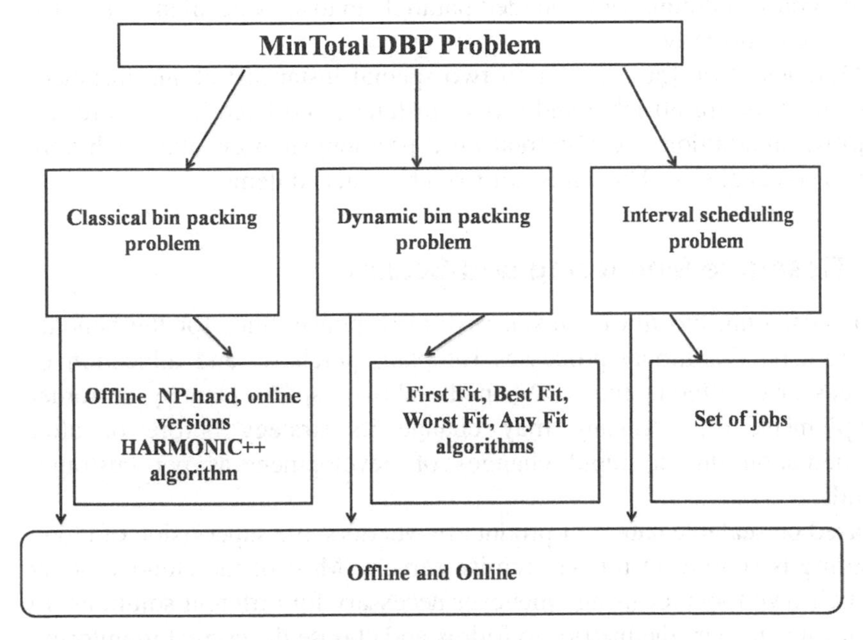

Figure 10.3 MinTotal DBP problems are involved with different problems.

easily through any online bin packing algorithms. Reader can refer [52] and [35] to determine both upper and lower bound detection.

(b) DBP problem: In this problem, items come in and go out randomly [41]. It mainly focuses to less or more number of bins utilized in the packing. Here several Fit packing algorithms are considered such as First Fit, Best Fit, Worst Fit, Any Fit algorithms, etc., with corresponding competitive ratios. Coffman *et al.* [41] used the First Fit packing algorithm with competitive ratio between 2.75 and 2.897. No online algorithm can obtain a competitive ratio less than 2.5 next to an optimal offline adversary. Chan *et al.* [38] demonstrated that competitive ratio with lower bound 2.5 holds without repacking by offline adversary. But in reverse, Ivkovic *et al.* [47] and Chan *et al.* [37] show less competitive ratio through online algorithm.

(c) Interval scheduling problem: The classical interval scheduling problem is defined through a set of jobs associated with a weight and run by interval scheduling[49]. Each machine can process only a single job at any time. Thus, a fixed number of machines maintains maximum weight as per possible jobs [36]. Flammini *et al.* [43] have considered extended interval scheduling with bounded parallelism to activate all machines for jobs concurrently.

Mertzios *et al.* [50] have used two special instances: clique instances (same time for all jobs) and proper instances (no fixed intervals for all jobs). In addition, the MinTotal DBP problem supposes that each item has no equal size. Thus, it is not fixed bin packed items.

10.6 Resource Monitoring and Scaling

Cloud computing has always taken several responsibilities for the benefits of customers. Traditional processes for plan, purchase, and allocation of resources take different times. When the above aspects plan by customer to implement, the company may change its strategy before or after implementation due to rapid changes of development as per customer demands.

Based on scalable nature of product by vendors, the supervision of cloud computing is significant for on-premise servers. Most of the cloud vendors have their own tools for using whenever necessary for different solutions. In this part, it considers the metrics to follow and choose the correct monitoring tools for requirements.

Cloud metrics:

Several cloud services are provided by cloud service providers with different metrics.Thus, it requires appropriate metrics to track and get tools for fulfilling customer needs. Cloud metrics is designed with sub-metrics as shown in Figure 10.4. In addition, the common metrics supply several supervising activities such as:

1. performance;
2. cost;
3. uptime.

(a) Performance:

The performance is determined through several factors that are provided by cloud vendors based on essential applications. But competitions for cloud services are always heated up in the market. It forces all vendors to pay more attention toward the performance of any applications. It focuses on supervising performance to see all metrics centrally. Again, based on hybrid cloud solutions, the monitoring tools always provide consistent reports. So,

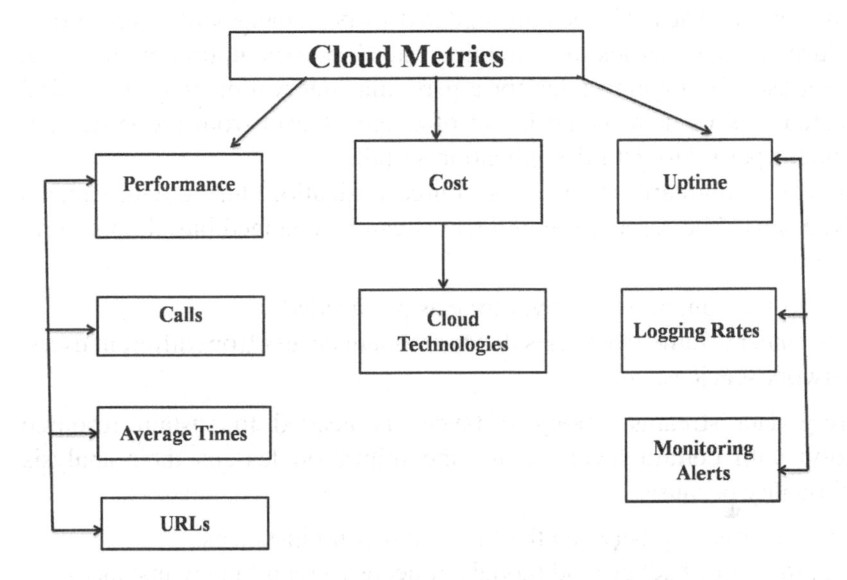

Figure 10.4 Cloud metrics designed with sub-metrics.

it always generates smart services as the customer demands. For example, recognizing slow queries needs performance monitoring. So, the significant metrics is needed to hold track per query such as:

- **Calls:** The numerous calls to a particular query.
- **Average time:** The average time taken for a particular query to run.
- **URLs:** The query is used through URLs in the web application.

(b) Costs:

Generally, establishing or using good cloud technologies save money. More money is needed as per allocation or uses of more resources. This use or allocation is elaborated during agreements with cloud vendors to activate their applications. To control the cost, it needs to supervise always the usage of resources; otherwise, the cost will be more automatic. Thus, costs depend on customer utilization of resources with highcaution.

(c) Uptime:

For failure architecture, correction is needed to be repaired as quickly as possible; otherwise, decision will be changed accordingly. It emphasizes the metrics for significant aspects of the system.

The above factors are very important for supervising or auditing through cloud metrics. Metrics are updated as per changes of applications. Particularly, it uses metrics like "logging rates" to assist uptime monitoring. If it supervises the logging rates for a particular interval of time, it can find unexpected lack of data which is out of range of data from the system. It subtracts the period to get the application's uptime.

It needs to regularly supervise resource utilization during data streams from NetCams. The resource requirements can be changed based on several reasons such as:

(i) The visual content varies over time as per needed.
(ii) The camera frame rate varies due to frequent access from different users, network services, etc.

Migrating data streams among instances is needed to sustain resource utilization with certain levels. Thus, the migration lessens their analysis results' quality because:

(i) There is time gap between the first and second instances.
(ii) Due to loss of background models, it needs to rebuild new instances.

To lessen the above effects, the resource manager decides to deallocate the under utilized instances. But for over-utilized instances, the manager supports migrating:

(a) Data streams for images are analyzed independently without loss of temporal information.
(b) Streams maintain low frame rate to lessen the consequence of time gap during the migration.
(c) Streams use resource intensive programs where fewer streams are migrated.
(d) Due to migrating of resource intensive with streams, streams try to maintain high resolutions for images.

10.7 Conclusion

The cloud resource manager lessens the cost over data streams from numerous NetCams to meet the performance requirements. The manager decides to allocate cloud instances through different factors such as analysis programs, needed frame rates, camera resolutions, the types and costs of the instances, etc. The manager generates the resource allocation problem as bin packing problem and finds its solution using a heuristic algorithm. The resource manager monitors, allocate, or deallocate instances as needed.

The MinTotal DBP problem focuses to lessen the whole cost of the bins used over time. It elaborated competitive ratios as per updated versions of the classic AnyFit algorithms. Several fit algorithms are used to determine better competitive ratio among them. It also finds gap between the existing upper and lowerbounds on the competitive ratios on different Fit. Again, it inspects the DBP problem to permit the assigned subset of bins providing the interactivity constraints in cloud gaming through distributed clouds.

References

[1] "Network camera and video analytics market," http://www.marketsandmarkets.com/Market-Reports/visual-communication-market-75.html.

[2] W.-T. Su, Y.-H. Lu, and A. S. Kaseb, "Harvest the information from multimedia big data in global camera networks," in Proceedings of the IEEE International Conference on Multimedia Big Data, , pp.184–191, 2015.

[3] A. S. Kaseb, E. Berry, E. Rozolis, K. McNulty, S. Bontrager, Y. Koh,Y.-H. Lu, and E. J. Delp, "An interactive web-based system using cloud for large-scale visual analytics," in Proceedings of the Imaging and Multimedia Analytics in a Web and Mobile World, 2015.

[4] N. Jacobs, N. Roman, and R. Pless, "Consistent temporal variations in many outdoor scenes," in Proceedings of the IEEE Conference on Computer Vision and Pattern Recognition, pp. 1–6, 2007.

[5] S. Srivastava and E. J. Delp, "Video-based real-time surveillance of vehicles," Journal of Electronic Imaging, vol. 22, no. 4, p. 041103,2013.

[6] Z. Chen, F. Yang, A. Lindner, G. Barrenetxea, and M. Vetterli,"Howis the weather: Automatic inference from images," in Proceedings of the IEEE International Conference on Image Processing, 2012, pp. 1853–1856, 2012.

[7] K. Hong, M. Voelz, V. Govindaraju, B. Jayaraman, and U. Ramachandran,"A distributed framework for spatio-temporal analysis on large-scale camera networks," in Proceedings of the IEEE International Conference on Distributed Computing Systems Workshops, pp. 309–314, 2013.

[8] K. Hong, S. Smaldone, J. Shin, D. Lillethun, L. Iftode, and U. Ramachandran,"Target container: A target-centric parallel programming abstraction for video-based surveillance," in Proceedings of the ACM/IEEE International Conference on Distributed Smart Cameras, pp. 1–8, 2011.

[9] X. Sun, N. Ansari, and R. Wang, "Optimizing resource utilization of a data center," IEEE Communications Surveys Tutorials, vol. 18,no. 4, pp. 2822–2846, 2016.

[10] X. Sun and N. Ansari, "Primal: Profit maximization avatar placement for mobile edge computing," in Proceedings of the IEEE International Conference on Communications, pp. 1–6, 2016.

[11] Xiang Sun, Nirwan Ansari, "Edgeiot: Mobile edge computing for the internet of things,"IEEE Communications Magazine, vol. 54, no. 12, pp. 22–29, 2016.

[12] U. Sharma, P. Shenoy, S. Sahu, and A. Shaikh, "A cost-aware elasticity provisioning system for the cloud," in Proceedings of the International Conference on Distributed Computing Systems, pp. 559–570June 2011.

[13] M. S. Hossain, M. M. Hassan, M. A. Qurishi, and A. Alghamdi, "Resource allocation for service composition in cloud-based video surveillance platform," in Proceedings of the IEEE International Conference on Multimedia and Expo Workshops, pp. 408–412, 2012.

[14] S. Vijayakumar, Q. Zhu, and G. Agrawal, "Dynamic resource provisioning for data streaming applications in a cloud environment,"in Proceedings of the IEEE International Conference on Cloud Computing Technology and Science, pp. 441–448, 2010.

[15] A. S. Kaseb, A. Mohan, and Y.-H. Lu, "Cloud resource management for image and video analysis of big data from network cameras," in Proceedings of the International Conference on Cloud Computing and Big Data, 2015.

[16] W. Chen, Y.-H. Lu, and T. J. Hacker, "Adaptive cloud resource allocation for analysing many video streams," in Proceedings of the IEEE International Conference on Cloud Computing Technology and Science, 2015.

[17] A. Mohan, A. S. Kaseb, Y. H. Lu, and T. J. Hacker, "Location based cloud resource management for analyzing real-time videos from globally distributed network cameras," in Proceedings of the IEEE International Conference on Cloud Computing Technology and Science, pp. 176–183, ,2016.

[18] D. S. Johnson, "Fast algorithms for bin packing," Journal of Computer and System Sciences, vol. 8, no. 3, pp. 272–314, 1974.

[19] M. R. Garey and D. S. Johnson, Computers and Intractability: A Guide to the Theory of NP-Completeness. W. H. Freeman & Co., 1979.

[20] D. K. Friesen and M. A. Langston, "Variable sized bin packing," SIAM journal on computing, vol. 15, no. 1, pp. 222–230, 1986.

[21] F. C. Spieksma, "A branch-and-bound algorithm for the two dimensional vector packing problem," Computers & operations research, vol. 21, no. 1, pp. 19–25, 1994.
[22] C. Chekuri and S. Khanna, "On multidimensional packing problems," SIAM journal on computing, vol. 33, no. 4, pp. 837–851, 2004.
[23] B. T. Han, G. Diehr, and J. S. Cook, "Multiple-type, two dimensional bin packing problems: Applications and algorithms,"Annals of Operations Research, vol. 50, no. 1, pp. 239–261, 1994.
[24] P. Kaew TraKul Pong and R. Bowden, "An improved adaptive background mixture model for real-time tracking with shadow detection," in Video-Based Surveillance Systems. Springer US, pp. 135–144, 2002.
[25] Z. Zivkovic, "Improved adaptive gaussian mixture model for background subtraction," in Proceedings of the International Conference on Pattern Recognition, vol. 2, pp. 28–31, 2004.
[26] J. Shi and C. Tomasi, "Good features to track," in Proceedings of the IEEE Conference on Computer Vision and Pattern Recognition, pp. 593–600, 1994.
[27] J.-Y. Bouguet, "Pyramidal implementation of the affine lucaskanade feature tracker description of the algorithm," Intel Corporation,vol. 5, no. 1-10, p. 4, 2001.
[28] N. Dalal and B. Triggs, "Histograms of oriented gradients for human detection," in Proceedings of the IEEE Conference on Computer Vision and Pattern Recognition, vol. 1, pp. 886–893, 2005.
[29] G. Bradski, "The OpenCV library," Dr. Dobb's Journal of Software Tools, vol. 25, no. 11, pp. 120, 122–125, 2000.
[30] W. Chen, A. Mohan, Y.-H. Lu, T. Hacker, W. T. Ooi, and E. J. Delp,"Analysis of large-scale distributed cameras using the cloud,"IEEE Cloud Computing, vol. 2, no. 5, pp. 54–62, 2015.
[31] Distribution and monetization strategies to increase revenues from cloud gaming. http://www.cgconfusa.com/report/documents/Content-5minCloudGamingReportHighlights.pdf.
[32] Gaikai. http://www.gaikai.com/.
[33] Onlive. http://www.onlive.com/.
[34] Streammygame. http://www.streammygame.com/smg/index.php.
[35] J. Balogh, J. Bt'ekt'esi, and G. Galambos. New lower bounds for certain classes of bin packing algorithms. Approximation and Online Algorithms(Lecture Notes in Computer Science, Volume 6534), pages 25–36, 2011.

[36] A. Bar-Noy, R. Bar-Yehuda, A. Freund, J. S. Naor, and B. Schieber. Aunified approach to approximating resource allocation and scheduling.Journal of the ACM, 48(5):735–744, Sept. 2001.

[37] J. W.-T. Chan, T.-W. Lam, and P. W. Wong. Dynamic bin packing of unit fractions items. *Theoretical Computer Science*, 409(3):521–529,2008.

[38] W.-T. Chan, P. W. Wong, and F. C. Yung. On dynamic bin packing: An improved lower bound and resource augmentation analysis. *Computingand Combinatorics*, pages 309–319, 2006.

[39] K.-T. Chen, Y.-C. Chang, P.-H. Tseng, C.-Y. Huang, and C.-L. Lei.Measuring the latency of cloud gaming systems. In *Proc. ACM MM'11*,pages 1269–1272. ACM, 2011.

[40] E. G. Coffman, J. Csirik, G. Galambos, S. Martello, and D. Vigo. Bin packing approximation algorithms: Survey and classification. Handbook of Combinatorial Optimization (second ed.), pages 455–531, 2013.

[41] E. G. Coffman, Jr, M. R. Garey, and D. S. Johnson. Dynamic binpacking. *SIAM Journal on Computing*, 12(2):227–258, 1983.

[42] E. G. Coffman, Jr., M. R. Garey, and D. S. Johnson. Approximation algorithms for bin packing: A survey. Approximation Algorithms for NP-hard Problems, pages 46–93, 1997.

[43] M. Flammini, G. Monaco, L. Moscardelli, H. Shachnai, M. Shalom,T. Tamir, and S. Zaks. Minimizing total busy time in parallel scheduling with application to optical networks. In *Proc. IEEE IPDPS'09*, pages1–12. IEEE, 2009.

[44] G. Galambos and G. J. Woeginger. On-line bin packing a restricted survey. *Zeitschriftflur Operations Research*, 42(1):25–45, 1995.

[45] M. R. Garey and D. S. Johnson. Computers and intractability: A guideto the theory of np-completeness, 1979.

[46] C.-Y. Huang, C.-H. Hsu, Y.-C. Chang, and K.-T. Chen. Gaming any where:an open cloud gaming system. In *Proc. ACM MMSys'13*, pages36–47. ACM, 2013.

[47] Z. Ivkovic and E. L. Lloyd. Fully dynamic algorithms for bin packing:Being (mostly) myopic helps. SIAM Journal on Computing, 28(2):574–611, 1998.

[48] J. W. Jiang, T. Lan, S. Ha, M. Chen, and M. Chiang. Joint vm placement and routing for data center traffic engineering. In *Proc. IEEEINFOCOM'12*, pages 2876–2880. IEEE, 2012.

[49] E. L. Lawler, J. K. Lenstra, A. H. RinnooyKan, and D. B. Shmoys.Sequencing and scheduling: Algorithms and complexity.

Handbooks in Operations Research and Management Science, 4:445–522, 1993.

[50] G. B. Mertzios, M. Shalom, A. Voloshin, P. W. Wong, and S. Zaks. Optimizing busy time on parallel machines. In *Proc. IEEE IPDPS'12*,pages 238–248. IEEE, 2012.

[51] P. E. Ross. Cloud computing's killer app: Gaming. IEEE Spectrum, 46(3):14–14, 2009.

[52] S. S. Seiden. On the online bin packing problem. Journal of the ACM, 49(5):640–671, 2002.

[53] C. Sharon, W. Bernard, G. Simon, C. Rosenberg, et al. The brewing storm in cloud gaming: A measurement study on cloud to end-userlatency. In Proc. ACM NetGames'12, 2012.

[54] A. Stolyar. An infinite server system with general packing constraints. *arXiv preprint arXiv:1205.4271*, 2012.

[55] C. Zhang, Z. Qi, J. Yao, M. Yu, and H. Guan. vgasa: Adaptive scheduling algorithm of virtualized gpu resource in cloud gaming. IEEETransactions *on Parallel and Distributed Systems*, accepted to appear,2014.

[56] Ahmed S. Kaseb, Anup Mohan, Youngsol Koh, Yung-Hsiang Lu, Cloud Resource Management for Analyzing Big Real-Time Visual Data from Network Cameras, IEEE Transactions on Cloud Computing, Volume: 7, Issue: 4, Page(s): 935 – 948, Oct.-Dec. 1 2019.

[57] Yusen Li, Xueyan Tang, Wentong Cai, Dynamic Bin Packing for On-Demand Cloud Resource Allocation, IEEE Transactions on Parallel and Distributed Systems, Volume: 27, Issue: 1, Page(s): 157 – 170, Jan. 1 2016.

11

Software-Defined Networking for Healthcare Internet of Things

Rinki Sharma

Ramaiah University of Applied Sciences
E-mail: rinki.cs.et@msruas.ac.in

Abstract

Internet of Things (IoT) has become an integral part of industry operation over the past few years. IoT applications have seen remarkable rise in the industries such as healthcare, agriculture, supply chain, smart energy, building and industrial automation, and connected cars. This chapter focuses on the healthcare industry. The chapter presents the applications and challenges of using IoT in healthcare industry and discusses the advantages, challenges, and possible solutions of incorporating software-defined networks (SDN) in healthcare Internet of Things (H-IoT). An architecture for SDN-based H-IoT system is presented and discussed. The chapter also discusses open research challenges, possible solutions, and research opportunities of integrating SDN with H-IoT.

Keywords: Internet of Things (IoT), healthcare Internet of Things (H-IoT), software-defined networking (SDN), healthcare.

11.1 Introduction

In the recent years, there has been tremendous rise in the number of Internet of Things (IoT) devices. IoT network comprises sensors, appliances, and other physical devices capable of communicating without human intervention. Wireless sensor networks (WSNs) are the building blocks of IoT. According to [1], there will be over 25 billion inter-connected IoT devices in 2020 with a rise of about 300 million new devices every month. IoT-enabled devices are transforming the healthcare industry by redefining the interaction between patients and medical professionals. IoT in healthcare benefits patients and their families, physicians, hospitals, and insurance companies. Wearable healthcare devices such as fitness bands, pulse oximeter, glucometer, blood pressure, and heart rate monitoring devices enable constant tracking of the health condition of a patient. Physicians can access a patient's health remotely and suggest appropriate possible treatment depending on patient's location [2]. Wearable devices for fitness monitoring allow people to keep a check on their fitness levels and calorie consumption. Connecting these devices to cloud allows users to monitor the variables over time and take appropriate actions or precautions. Hospitals use IoT not only to monitor patient's health and location but also to track real-time location and functioning of medical equipment tagged with sensors.

Analysts anticipate that, by 2020, the data generated by IoT devices will be about 270 times higher than the data generated by human operated end-user devices, and the connected IoT devices will generate 79.4 ZB of data in 2025 [3]. This rise of the communicating IoT devices leads to exponential rise of network traffic, thus requiring high network bandwidth, computation, and storage. IoT devices mostly communicate over wireless medium and comprise mobile nodes. As the number of such devices increases in a given network area, network performance drops due to interference among participating nodes and lack of sufficient network bandwidth [4]. Changing network patterns call for the need to change the traditional networking methods. To support healthcare IoT (H-IoT), the transmission networks should be capable of supporting concurrent multiservice data flows with effective isolation and prioritization. Network configuration should be flexible and scalable while utilizing the network bandwidth effectively. This is where software-defined networking (SDN) plays a crucial role. SDN allows the service providers to dynamically allocate the bandwidth and resources based on the requirements by the application [5]. Unlike traditional networks that depend on hardware routers and switches for network operation and connectivity, SDN separates the data and control planes and uses the network controller to make necessary decisions for dynamic allocation of network resources based on the network traffic. With increase of mobile end users, exponential increase in network data generation, bandwidth intensive applications, and heterogeneous network architectures, SDN is capable of efficiently adapting to changing network requirements [6]. This ability of SDN is crucial for H-IoT applications, in particular, because these applications demand high bandwidth and low latency.

This chapter presents the significance of SDN in H-IoT. Section 11.2 discusses the applications, communication technologies, characteristics, and challenges for H-IoT networks. Section 11.3 presents a brief introduction to SDN along with its advantages. The architecture of SDN-based H-IoT systems is discussed. Section 11.4 presents the open research challenges and opportunities of integrating SDN with H-IoT. Section 11.5 concludes the chapter. The references are presented in Section 11.6.

11.2 Healthcare Internet of Things

The idea of connected devices has been around since the 1970s. Back then, it was called "embedded internet" or "pervasive computing." In 1999, Kevin Ashton coined the term "Internet of Things" for the radio frequency identification (RFID) technology [7]. The concept of IoT started to gain popularity around 2011 with advancement of Internet protocol version 6 (IPv6). With 128-bits-long IPv6 and the availability of 2^{128} unique IP addresses, various devices and appliances could have an IP address as a unique ID and networked to participate in communication over Internet. Over the years, IoT has found applications in the following areas:

1. Healthcare
2. Connected industry
3. Smart city
4. Smart energy
5. Connected car
6. Smart agriculture
7. Connected building
8. Retail
9. Supply chain

This chapter concentrates on the healthcare applications of IoT. H-IoT is gaining popularity to provide remote healthcare services such as diagnosing, monitoring, and remote surgeries over the Internet [8]. Smart rehabilitation is used to take care of the aged. H-IoT connects healthcare resources and assistive devices with doctors, nurses, care takers, hospitals, and rehabilitation centers. Some of the popular H-IoT applications as presented in Figure 11.1 are as follows:

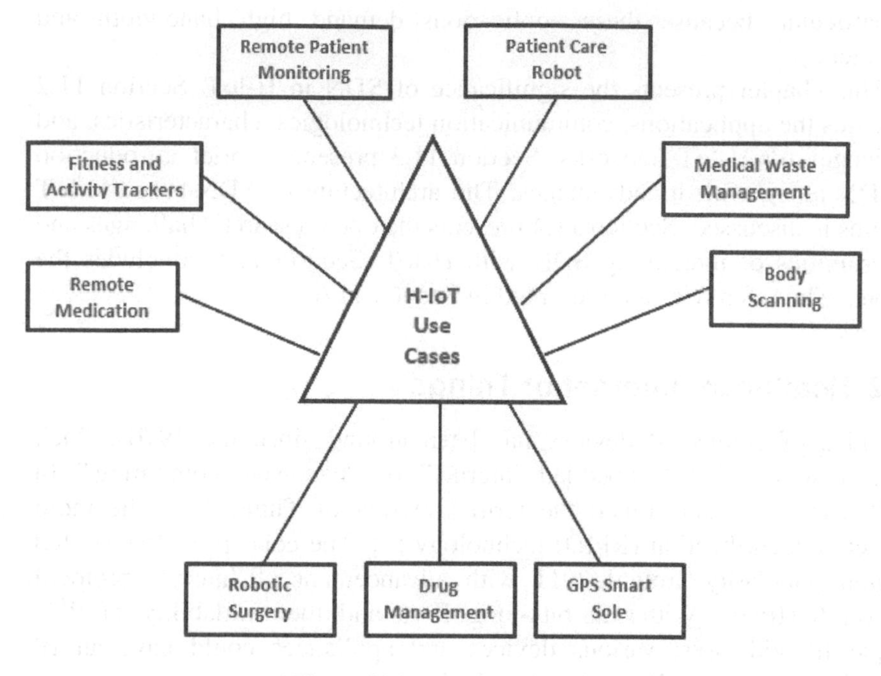

Figure 11.1 H-IoT use cases.

1. **Remote patient monitoring:** Wearable devices and sensors are used to monitor the vital parameters of the patient. This information can be transmitted wirelessly to the doctor's/nurse's device or to the cloud from where it can be read by the medical staff involved in taking care of the patient. Such a facility allows the patient to move around instead of sitting at one place and allows doctors to monitor patient's vitals without requiring to visit the patient often [2].
2. **Fitness and activity trackers:** People wearing fitness and activity trackers mounted on wrist, ankle, or belt is a common sight these days. These trackers check the physical activity of the user such as steps taken and stairs climbed, and vital health parameters such as pulse, heart-beat, body temperature, and other basic physiological parameters. This data is stored on the cloud and users can study it overtime to track their fitness and activity levels [9].
3. **Remote medication through ingestible sensors:** Ingestible sensors coated with magnesium and copper are used to trigger a signal if medication is not taken by a patient within appropriate time [10].

4. **Robotic surgery:** To perform surgical operations with enhanced control and precision, surgeons use IoT-enabled robotic surgery devices. Surgeons can also perform surgery through remote locations using these devices. However, this calls for real-time response over a high bandwidth communication channel [11].
5. **Drug management:** Drug supply, distribution, management, and monitoring are carried out by using RFID. This helps in achieving enhanced supply chain management and reduced overall production costs. In case of wrong distribution, reports can be sent to the regulatory bodies through cloud [12].
6. **GPS smart sole:** A smart insole embedded with global positioning system (GPS) can be used to monitor the movement and location of the patient. This application of H-IoT can be particularly useful to monitor the movement of aged people or Alzheimer patients [13].
7. **Body scanning:** Smart body scanner is used for regular fitness checks as well as to check the vitals of a person. A weight scale, also known as the turntable, spins the body of the person concerned for 15 seconds. 3D cameras are used to capture 360° image of the person to develop a 3D model of the person's body. A body fat, circumference, vitals, and weight scan is sent on the mobile devices. A comparison of new readings with old values of the parameters is also made available [14].
8. **Medical waste management:** Medical waste disposal, monitoring, and management are critical issues in healthcare. Automated medical waste management uses sensors embedded in dustbins located at different locations in the hospital to notify the quantity of waste. When the waste needs to be cleared, an automated robot is used to collect the waste and dump it in an appropriate place. This avoids any human interaction with hazardous medical waste and keeps the environment clean [15].
9. **Patient care robot:** These robots are used to provide care and support to the patients in the hospital. They are used to deliver medication, food, and necessary items to the patient. Equipped with on-board sensors, these robots navigate using built-in maps and communicate over wireless medium [16].

The communication technologies used for radio communication can be classified into long and short distance communication technologies. The long distance communication technologies involve technologies such as Internet, mobile communication, and cloud. The short distance radio communication technologies used for H-IoT are presented in Table 11.1.

Table 11.1 Short distance radio communication technologies for H-IoT.

	Wi-Fi	Zigbee	Bluetooth	RFID	IrDA	UWB
International Standard	IEEE 802.11a/b/g/n	IEEE 802.15.4	IEEE 802.15.1	ISO 18047	–	IEEE 802.15.3a
Data rate	1–450 Mbps	20–250 kbps	Up to 3 Mbps	106–424 kbps	14.4 kbps	53–480 Mbps
Frequency band of operation	2.4 GHz, 5 GHz	868 MHz, 2.4 GHz	2.4 GHz	13.56 MHz	850–900 nm	3.1–10.6 GHz
Communication distance	200 m	10–20 m	20–200 m	20 cm	0–1 m	0–10 m
Communication type	Point-to-multipoint	Point-to-multipoint	Point-to-point	Point-to-point	Point-to-point	Point-to-point
Security	SSID, WEP, WPA	AES	AES, ECDH	AES	–	AES
Power (mW)	>1000 mW	100 mW	1–100mW	<1 mW	1 μW	<1 mW

These technologies vary in their data rates, communication range, frequency of operation, transmission power, communication power, and security. Based on the application requirement and services provided by the communication technologies, appropriate technology is used for H-IoT.

The requirements of H-IoT cannot be fulfilled by a single communication technology. The requirements of the H-IoT applications may vary in terms of communication range, data rate (communication speed), spectrum (licensed vs. unlicensed), and transmission power.

By using wireless fidelity (Wi-Fi) technologies, the medical and IoT devices all over the campus can be connected for reliable communication by virtue of its communication range and supported data rate. Wi-Fi supports device mobility over the network, thus providing dynamic network environment.

Zigbee is a popular WSN technology that uses unlicensed wireless spectrum. Various medical sensors make use of Zigbee for communication. In H-IoT, the most common use of Zigbee is in remote patient monitoring. Body sensor network used for patient health monitoring, fitness, and activity trackers make use of Zigbee sensors. Zigbee provides an appropriate balance of parameters such as data rate, communication range, and transmission power to make it the chosen technology for these applications.

Bluetooth technology supports both voice and data, provides 1 Mbps of data rate, and uses unlicensed spectrum for communication. These features of Bluetooth make it a suitable choice for H-IoT. In a hospital campus, Bluetooth devices can be identified and visitors can be informed about the floor plan, doctors on duty, and appointment scheduling through Bluetooth beacons. Bluetooth can be used for automated patient check-in and check-out, optimized patient flow, scheduling appointments with doctors, compliance tracking and recording, asset tracking, and wayfinding.

RFID is a short distance, non-contact communication technology, which does not require direct line of sight communication. It is an economical and reliable technology used for location and identification of objects. In H-IoT applications, RFID is extensively used to identify, locate, and track medical equipment.

Infrared Data Association (IrDA) is wireless optical communication technology with a very short distance, point-to-point, line of sight and very low bit rate. This is used to carry out physically secured data transmission. Infrared (IR) technology is used in remote control of appliances such as television, projector, air-conditioner, and such other devices. In healthcare, IrDA can be used for remotely controlling different medical devices.

Ultra-wideband (UWB) is a short distance communication technology that provides data rates of up to 450 Mbps. This uses low-power pulses for communication and generates low electromagnetic radiation, thus making it suitable for medical applications. UWB radar is used for monitoring of patient's motion over short distances. Real-time exchange of medical images (such as X-rays-, cardiology-, pneumology-, obstetrics-, and ENT-imaging) and data over a distance of up to 10 m.

There are three main actions involved in H-IoT system as shown in Figure 11.2. First, the sensors or devices (things equipped with sensors) acquire the concerned data. Some of these devices are capable of pre-processing the data before transmission.

The H-IoT should be able to acquire accurate patient data and transmit/update it in real time. This requires reliable low-latency communication. The characteristics of H-IoT are low latency, low power, high reliability, and secure operation. H-IoT systems are time-critical in nature; therefore, the end-to-end transmission delays need to be minimal and network quality of service (QoS) needs to be high. Data security and privacy are essential for H-IoT communication as it involves exchange of

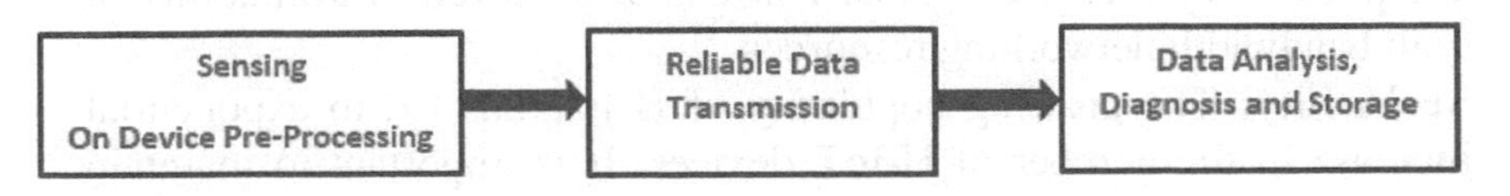

Figure 11.2 Components of an H-IoT system.

patient data and information regarding vital parameters over the network. The sensors involved in H-IoT communication are resource constrained with limited battery and processing power. Therefore, the communication protocols and technologies used for H-IoT should be lightweight and require low transmission power. Different long-distance and short-distance wireless communication technologies are used for this. Some of the popularly used short-distance wireless communication technologies are presented in Table 11.1. When the data reaches the receiving end, it is analyzed and used for diagnosis. Big data analytics plays a great role at the receiving stations for fast, efficient, and reliable visualization, exploration and analysis of the received data. The decision on patient diagnosis can be taken based on the obtained analysis. The received data is stored in appropriate servers or cloud for further analysis or processing.

11.2.1 Challenges in H-IoT

While H-IoT applications are becoming immensely popular in the healthcare industry, numerous challenges are faced in their implementation. This section presents some of the crucial challenges in H-IoT implementation.

1. **Heterogeneous networks:** Implementation of H-IoT involves a variety of networks and requires devices implemented on varying hardware, operating systems, and programming languages to coexist and interoperate. The communication technologies used are short-range such as Bluetooth, IrDA, UWB and Zigbee, as well as long-range such as Wi-Fi and global system for mobile communication (GSM) and/or long-term evaluation (LTE). Long-range communication leads to the device battery draining out faster. It is necessary that despite of these heterogeneities in implementation and communication, these devices should be able to inter-operate by identifying and discovering each other.
2. **Quality of service:** H-IoT is characterized by low-latency, low-power operation with high reliability. As H-IoT deals with critical issues involving a patient's health, they are bound to be time-critical by nature. Therefore, minimal delays should be incurred in transmission and processing of H-IoT signals. This can be achieved by availability of high bandwidth networking resources.
3. **Scalability:** The growing popularity of H-IoT has led to exponential increase in the number of H-IoT devices. It is important to maintain consistent performance and QoS in the H-IoT network even as the new devices are added in the network. To support large-scale deployment of

H-IoT networks, this issue needs to be addressed effectively to avoid system downtime particularly in crucial H-IoT networks.

4. **Security and privacy:** Security and privacy of the patients and their data is of paramount importance in H-IoT. In many cases, H-IoT data is communicated for long distances over heterogeneous networks and stored in the cloud, making it vulnerable to attacks. Therefore, robust encryption and authentication techniques for security and privacy of data are essential. However, while designing these techniques, it has to be taken into consideration that the devices used for H-IoT are resource constrained; hence, the developed solutions too should be lightweight and energy efficient.

11.3 Software-Defined Networking

The traditional networking architecture requires manual configuration of networking devices such as switches and routers. This traditional method of networking is deemed incapable of handling the ever-increasing demands for high-network QoS and performance because of the emergence of new traffic patterns due to growing popularity of applications in IoT, machine-to-machine (M2M), mobile and edge computing, escalation of 5G traffic, evolving wide area network (WAN) requirements, and other Internet-based applications [17]. Changing traffic patterns over the network due to node mobility require flexibility in network resource allocation while maintaining network QoS and performance up to the required standards. To encounter the constantly changing network demands SDN is being adapted by many networking platforms and organizations. Unlike in traditional networks, where network devices such as routers and switches need to be configured manually, in SDN, the network controller can be programmed and configured to respond to changes in the communication network such as traffic load and network topology.

SDN separates the control plane (the decision-making plane, mostly implemented in software) from the data plane (the data transfer plane, mostly implemented in hardware). The application plane comprises the applications configured by the network operator such as load balancer, access control and bandwidth allocation [18]. The SDN architecture is presented in Figure 11.3. The SDN network controller has the global view of the network and is capable of making decisions about resource allocation based on changing network requirements. The traffic forwarding policies are defined by the network operator while the centralized network controller facilitates automated

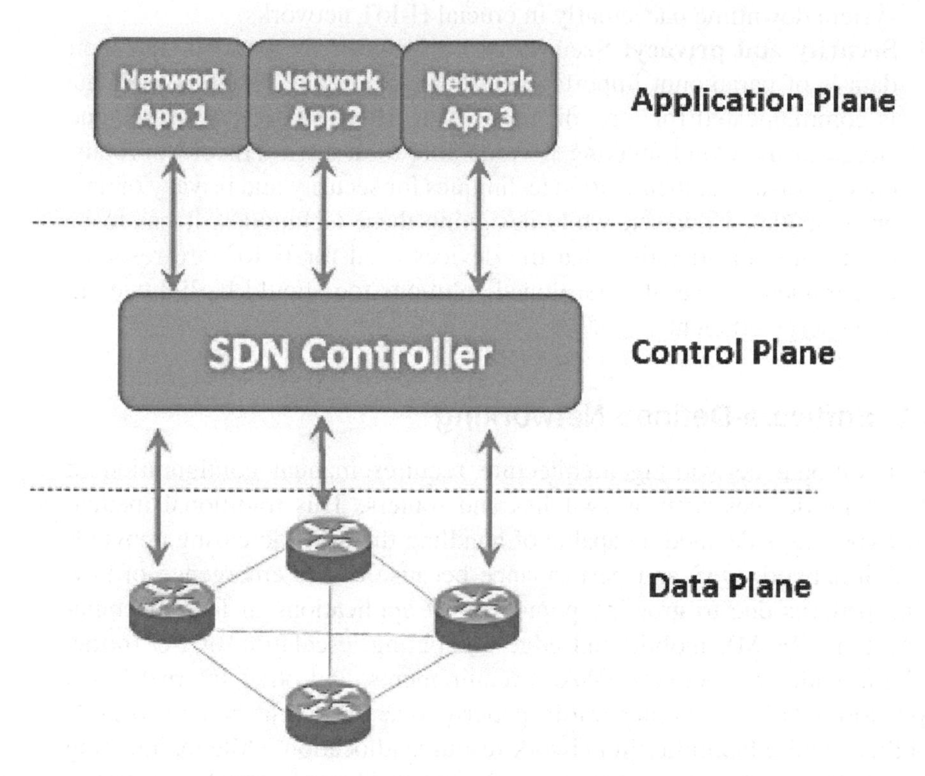

Figure 11.3 Typical data center infrastructure.

network management. To handle the increasing network demand, network resources can be virtualized. Network function virtualization (NFV) in association with SDN brings in numerous advantages to network performance while achieving reduced capital expenditure (CAPEX) and operational expenditure (OPEX) [19, 20]. The communication between network devices and applications takes place through the network controller. As the networks scale up, the number of network controllers incorporated into the network could be increased to maintain the network performance with increasing traffic load.

Some of the advantages brought in by SDN are:

1. **Dynamic resource allocation:** Network programmability facilitates programmable network infrastructure. This allows allocation of

bandwidth and other network resources dynamically based on changing network role.

2. **Reduced CAPEX and OPEX:** Network businesses can optimize their network hardware and deploy cost-efficient solutions. SDN also compliments NFV and cloud computing. With access to virtualized resources, organizations do not need to deploy physical resources at their end, thus attaining reduced CAPEX and OPEX.
3. **Efficient network management:** Network operators are allowed to define policies in line with business requirements of the organizations, which makes it easier to deploy and maintain the network. This allows SDN to be a key policy based automation which eliminates operational overhead and complexity particularly in highly scalable networks.
4. **Network centralization and management:** SDN allows a centralized view of the entire configured network at the network controller. This allows the network operators to manage the network configuration from one point. Management of physical and virtual network devices from a centralized controller offers ease and efficiency of dynamic network configuration based on network traffic demand which is not possible to achieve through traditional networks.
5. **Multivendor support:** The characteristic of SDN paradigm to automate network infrastructure requires it to be compatible with large multivendor ecosystem across a large number of infrastructure device types. SDN multivendor support provides the network operators with pronounced flexibility in network configuration.

An example of SDN-based H-IoT system is presented in Figure 11.4. The system is divided into the following three planes:

1. **Data plane:**
   The data plane comprises the network devices and infrastructure used to forward the data over the network. This plane is also called the forwarding plane. The network routers and switches fall in this plane. As shown in Figure 11.4, the data is acquired from sensors and other IoT devices through the wireless/wired network infrastructure. The received data is routed and forwarded to the network controller (in the control plane) through the data plane.
2. **Control plane:**
   Control plane is the decision-making plane. The main entity of this plane is the network controller that comprises the logic written for traffic

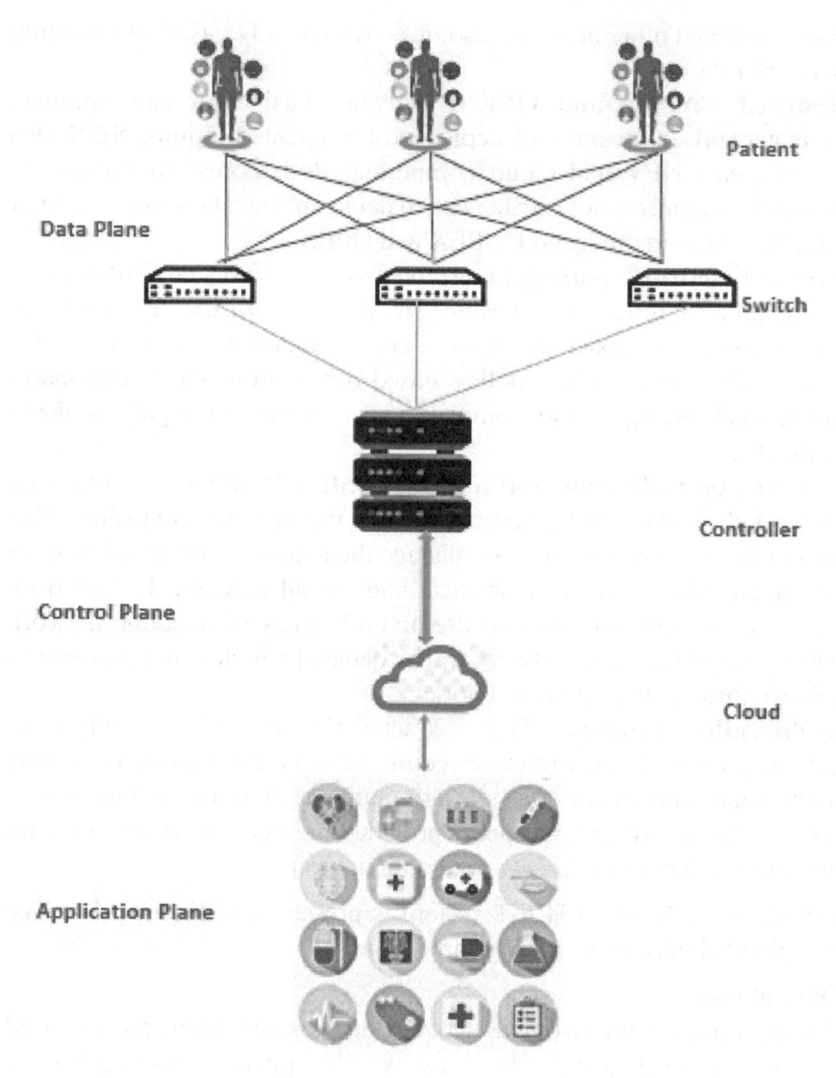

Figure 11.4 SDN-based H-IoT system.

control such as routing, load balancing, and traffic engineering. The controller fetches and maintains different network information such as network state, topology, and statistics and makes network infrastructure control decisions based on this information. The network controller receives application requirements from the application layer and makes decision regarding QoS and dynamic bandwidth allocation. It has an abstract view of the network, the events in the network, and network statistics.

The SDN can be combined with the cloud for better network management and network data access. This removes the barrier associated with network hardware access and builds the network implementation cost-effective and agile. As seen from Figure 11.4, the data from the network controller is uploaded on to the cloud. Similarly, the network controller also fetches the data from the cloud and transmits it to the end devices.

3. **Application plane:**

   The application plane presented in Figure 11.4 comprises all the healthcare-related applications that work with the H-IoT data. The application plane comprises the receiving applications that obtain the data originated at the patient. This data is used for diagnosis, analytics, generating statistics, and making appropriate decisions. As seen from Figure 11.4, the decisions made or messages can be conveyed back to the patient through the cloud and network controller.

11.4 Opportunities, challenges, and possible solutions for SDN-based H-IoT Networks

As the IoT applications are becoming increasingly popular in the healthcare industry, the number of H-IoT systems and the data exchanged over these networks is increasing constantly. As for many of H-IoT applications, the end nodes can be mobile. Dynamicity of these networks in terms of node mobility, required bandwidth, and on-demand resource allocation can be handled well by SDN. However, while SDN offers numerous advantages to H-IoT networks, there are certain challenges in the incorporation of SDN for H-IoT. This section discusses the opportunities, challenges, and possible solutions for successful integration of SDN in H-IoT.

Some of the opportunities of integrating SDN with H-IoT are as follows:

1. **Dynamic resource allocation:** H-IoT applications demand high QoS, high reliability, and low latency. Particular applications such as remote robotic surgery require real-time performance. Many of the H-IoT end devices could be mobile, leading to changing network topologies and bandwidth requirements. This call for the need to frequently adapt to the network changes and dynamically allocate appropriate network resources to support the required QoS by the H-IoT applications.

2. **Centralized network monitoring and flow management:** The SDN network controller acquires a global view of the entire network. This information enables the network controller to make instantaneous decisions on network traffic flow management, such as diverting traffic flow to a better available path to avoid network congestion, or allocate more network resources to achieve better network performance.
3. **Dynamic policymaking:** As H-IoT applications demand for high reliability and robustness, to support real-time data requirements for H-IOT applications, the SDN controller can be programmed to change policies dynamically to support the transfer of high priority packets reliably over the network.
4. **Network security:** As the SDN network controller processes the network traffic, it can implement intelligent and agile threat mitigation logic against hostile traffic flows.

While incorporation of SDN with H-IoT brings in numerous advantages to the H-IoT networks, the very characteristics of SDN also impose certain challenges to H-IoT networks. Some of these significant challenges and their possible solutions are as follows:

Challenge 1:

Centralized controller and reliability: The centralized control architecture of SDN uses the centralized network controller to maintain the network performance and for network management. This centralized network controller can become a single point of failure. High network traffic leads to high traffic load on the network controller, thus making it a bottleneck and leading to congestion. Bottleneck and congestion impacts the performance, QoS, and reliability of the network. Therefore, it is essential to overcome these issues by incorporating appropriate solutions.

Solution: Instead of relying on a single network controller, a network can comprise multiple redundant network controllers. In case the primary network controller fails, the secondary controller should take the control without any downtime [21].

Challenge 2:

Scalability: With increasing popularity of H-IoT, the number of devices in the H-IoT network and, hence, the traffic over these networks is increasing exponentially. For every new incoming packet, the SDN controller creates a flow table. As the network traffic increases, the computational overhead at the network controller also increases, leading to processing delays. This further leads to bottleneck and network congestion in the network, thus deteriorating network performance.

Solution: Load balancing over multiple network controllers can be used to overcome this challenge. Load balancing allows simultaneous processing of requests in SDN-based networks [22]. By using multiple network controllers, the data processing load can be distributed among these controllers, thus reducing the chances of bottleneck and network performance deterioration. The number of network controllers to be incorporated into the network and their placement can depend on the size of the network and the network traffic that needs to be handled by the controller.

Challenge 3:

Security: Like any other network, H-IoT networks too are prone to security attacks. SDN-based H-IoT comprises the network controller which can become the single point of attack to bring down the whole network. Hence, securing the controller is most crucial in securing SDN-based H-IoT network.

Solution: To protect the SDN network controller from malicious applications and traffic, solutions that could provide access control to the controller dynamically are required. It is also essential to implement solutions for secure control communication layer, authentication, authorization, and accounting (AAA), application program interface (API) security, logging/security audit service, and resource monitoring at the network controller [23].

11.5 Conclusion

This chapter presents the significance and role of SDN, integrated with H-IoT. Growing popularity of H-IoT has led to exponential increase in the number of H-IoT devices and H-IoT traffic over the network. Due to the presence of mobile nodes in the H-IoT networks, the network traffic in these networks is highly dynamic. As traditional network architecture is not suitable to handle such traffic, SDN plays an important role in these networks. The characteristics of SDN that make it a suitable choice for H-IoT networks are presented. The proposed architecture for an SDN-based H-IoT system is presented and discussed. While SDN is a good choice for H-IoT networks, they pose numerous challenges as well. The challenges posed by SDN for H-IoT networks are presented along with the possible solutions and research opportunities to overcome these challenges.

References

[1] http://www.syscomm.co.uk/iot/the-rise-of-iot-what-it-means-for-the-demand-on-our-networks/ accessed on 3 June 2020

[2] Sharma, R., Gupta, S.K., Suhas, K.K. and Kashyap, G.S., 2014, April. Performance analysis of Zigbee based wireless sensor network for remote patient monitoring. In 2014 Fourth International Conference on Communication Systems and Network Technologies (pp. 58-62). IEEE.

[3] https://www.idc.com/getdoc.jsp?containerId=prUS45213219

[4] Sharama, R., Shankar, J.U. and Rajan, S.T., 2014, April. Effect of Number of Active Nodes and Inter-node Distance on the Performance of Wireless Sensor Networks. In 2014 Fourth International Conference on Communication Systems and Network Technologies (pp. 69-73). IEEE.

[5] McKeown, N., 2009. Software-defined networking. INFOCOM keynote talk, 17(2), pp.30-32.

[6] Yeganeh, S.H., Tootoonchian, A. and Ganjali, Y., 2013. On scalability of software-defined networking. IEEE Communications Magazine, 51(2), pp.136-141.

[7] Ashton, K., 2009. That 'internet of things' thing. RFID journal, 22(7), pp.97-114.

[8] Tarouco, L.M.R., Bertholdo, L.M., Granville, L.Z., Arbiza, L.M.R., Carbone, F., Marotta, M. and De Santanna, J.J.C., 2012, June. Internet of Things in healthcare: Interoperatibility and security issues. In 2012 IEEE international conference on communications (ICC) (pp. 6121-6125). IEEE.

[9] Lomotey, R.K., Pry, J. and Sriramoju, S., 2017. Wearable IoT data stream traceability in a distributed health information system. Pervasive and Mobile Computing, 40, pp.692-707.

[10] Philip, V., Suman, V.K., Menon, V.G. and Dhanya, K.A., 2017. A review on latest internet of things based healthcare applications. International Journal of Computer Science and Information Security, 15(1), p.248.

[11] Shabana, N. and Velmathi, G., 2018. Advanced Tele-surgery with IoT Approach. In Intelligent Embedded Systems (pp. 17-24). Springer, Singapore.

[12] Liu, L. and Jia, W., 2010, September. Business model for drug supply chain based on the internet of things. In 2010 2nd IEEE InternationalConference on Network Infrastructure and Digital Content (pp. 982-986). IEEE.

[13] Wilden, J., Chandrakar, A., Ashok, A. and Prasad, N., 2017, October. IoT based wearable smart insole. In 2017 Global Wireless Summit (GWS) (pp. 186-192). IEEE.
[14] Muneer, A., Fati, S.M. and Fuddah, S., 2020. Smart health monitoring system using IoT based smart fitness mirror. Telkomnika, 18(1), pp.317-331.
[15] Raundale, P., Gadagi, S. and Acharya, C., 2017, July. IoT based biomedical waste classification, quantification and management. In 2017 International Conference on Computing Methodologies and Communication (ICCMC) (pp. 487-490). IEEE.
[16] Patel, A.R., Patel, R.S., Singh, N.M. and Kazi, F.S., 2017. Vitality of Robotics in Healthcare Industry: An Internet of Things (IoT) Perspective. In Internet of Things and Big Data Technologies for Next Generation Healthcare (pp. 91-109). Springer, Cham.
[17] Wan, J., Tang, S., Shu, Z., Li, D., Wang, S., Imran, M. and Vasilakos, A.V., 2016. Software-defined industrial internet of things in the context of industry 4.0. IEEE Sensors Journal, 16(20), pp.7373-7380.
[18] Shin, M.K., Nam, K.H. and Kim, H.J., 2012, October. Software-defined networking (SDN): A reference architecture and open APIs. In 2012 International Conference on ICT Convergence (ICTC) (pp. 360-361). IEEE.
[19] Ananth, M.D. and Sharma, R., 2017, January. Cost and performance analysis of network function virtualization based cloud systems. In 2017 IEEE 7th International Advance Computing Conference (IACC) (pp. 70-74). IEEE.
[20] Ananth, M.D. and Sharma, R., 2016, December. Cloud Management using Network Function Virtualization to Reduce CAPEX and OPEX. In 2016 8th International Conference on Computational Intelligence and Communication Networks (CICN) (pp. 43-47). IEEE.
[21] Kuroki, K., Fukushima, M., Hayashi, M. and Matsumoto, N., 2014. Redundancy method for highly available OpenFlow controller. International Journal on Advances in Internet Technology, 7(1).
[22] Sharma, R. and Reddy, H., 2019, December. Effect of Load Balancer on Software-Defined Networking (SDN) based Cloud. In 2019 IEEE 16th India Council International Conference (INDICON) (pp. 1-4). IEEE.
[23] Dangovas, V. and Kuliesius, F., 2014, January. SDN-driven authentication and access control system. In The International Conference on Digital Information, Networking, and Wireless Communications (DINWC2014) (pp. 20-23).

12

Cloud Computing in the Public Sector: A Study

Amita Verma[1] **and Anukampa**[2]

[1]Associate Professor, University Institute of Legal Studies, Panjab University, Chandigarh, India
[2]Research scholar, Kurukshetra University, Haryana, India.
E-mail: amitaverma21@gmail.com, anukampa001@gmail.com

Abstract

Clouds floating in the wide blue sky – some clouds are so huge that theyextend farther than our eye's reach, while some clouds are specifically the ones that cause rain, while some are small and fluffy and look like different shapes we loved to identify in our childhood. But, who thought technology would move at such a fast pace that one day we will have the "cloud" storing all our data, in consonance with our needs and be at our service. It was only a decade ago that mobile phones had memory cards to store the data on our phone including songs, images, videos, etc., but now we have the cloud for that. The need to download stuff for our instantaneous use onto our physical memory drive has been surpassed, to a certain extent, by the cloud. You want to listen to a song, open Spotify or YouTube or any other platform; they have a huge repository of pretty much all the songs, name it and it is there! Your client e-mailed you an important document, you do not need to download it onto the physical memory, you can access it from anywhere in the world on your Gmail account. These are small examples of cloud computing, the new age technology set to take the world by storm. This paper discusses in detail what exactly cloud computing is, its evolution, and how this brilliant technology can be used in the various agencies of the public sector along with the advantages of using it, which include easy sharing between different agencies, cost cutting, etc. and the challenges that come with it.

Keywords: Cloud computing, information technology, government, public sector.

12.1 Introduction

Ever struggled with a limited phone memory or free space, while trying to download something onto your phone? The struggle is real!

But the massive advancement that science has seen in the past few decades has successfully solved this problem too. The introduction of "cloud computing" works wonders when it comes to data storage, its management, and processing. So, now, you can store your data on the "cloud" instead of saving it on the hard drive of your computer, phone, or any other device.

Simply put, "cloud computing" is the delivery of various services on demand, through the Internet. The smallest and most widely used example of cloud computing is "Gmail." Google mail or Gmail stores all your e-mails in their remote servers (cloud) and the same can be accessed by you irrespective of your geographical location.

After the pandemic hit, most of the companies have adopted the "work from home" mode; the employees, irrespective of where they are, have access to the company data and files, relevant to the department they work in and the only requirement to access the same is an Internet connection.

Cloud computing has definitely made work easy as the user, in a way, shifts his responsibility and costs onto another, i.e., the cloud service provider in exchange for a nominal fee charged by the cloud service provider.

The user pays the nominal fee and primarily cuts his own individual costs that he would have recurred on getting more storage space onto his devices, or getting more devices and also hiring more staff to handle the same.

Four types of cloud computing systems exist – public, private, community, and hybrid. Public clouds cater to the needs of anybody over the Internet for free or for a small fee. A private cloud caters to a network of specific persons or an organization protected by a firewall. A community cloud is shared by more than one organization with similar needs and the combination of the two or more than two types of clouds also exists and is called a hybrid cloud.

12.2 History and Evolution of Cloud Computing

The origin of the concept of cloud computing is not as novel as it seems, rather some of the concepts were in existence since as early as the 1950s when mainframe computing was in and an organization only needed one to two computers for the entire staff operating therein, refer to Figure 12.1 Further, in 1969, J. C. R. Licklinder conceptualized and helped develop a very simple and one of the earliest systems of interconnected computers via the Internet for the purpose of sharing information called ARPANET, i.e., Advanced Research Projects Agency Network. After this, there is no looking back.[1]

With technological advancement and development moving further at a fast pace, Amazon Web Services and Google Docs were launched by Amazon and Google, respectively. While Amazon Web Services provided the users to access their programs and applications on rented computers, Google Docs could be utilized to store, edit, and share documents and files by the users.

Netflix, a life-saver and super popular online video streaming service, was not the same when it was established (1997) and launched its website in 1998. As egregious as it sounds, back then, Netflix mailed the DVDs, on demand, for a fee. However, now you search a title and Netflix has it, play, and enjoy your movie. The current model of Netflix has been in use since the year 2007, when it switched its modus operandi from mailing DVDs to storing it on the cloud and the users could view them online on the app or browser, a rather huge advancement in the journey of cloud computing.

As the shift to cloud computing started gaining momentum, various platforms started offering "*pay-as-you-go*" services like Software as a Service (SaaS), Platform as a Service (PaaS), Infrastructure as a Service (IaaS), Backup as a Service (BaaS), and so on and so forth, which have now become an indispensable part in the structure of an organization.

In a research by "451 Research"[2] on cloud computing it made a prescient observation that the revenue from cloud computing as an industry from $28.1 billion in the year 2017 will sky rocket to $53.3 billion in 2021.[3]

[1] https://the-report.cloud/the-evolution-of-cloud-computing-wheres-it-going-next accessed on 15/08/20 at 19:30

[2] A global IT analyst firm, part of S&P Global Market Intelligence.

[3] https://i1.wp.com/the-report.cloud/wp-content/uploads/2018/12/2-1.png?w=1024&ssl=1 accessed on 16/08/20 at 11:00

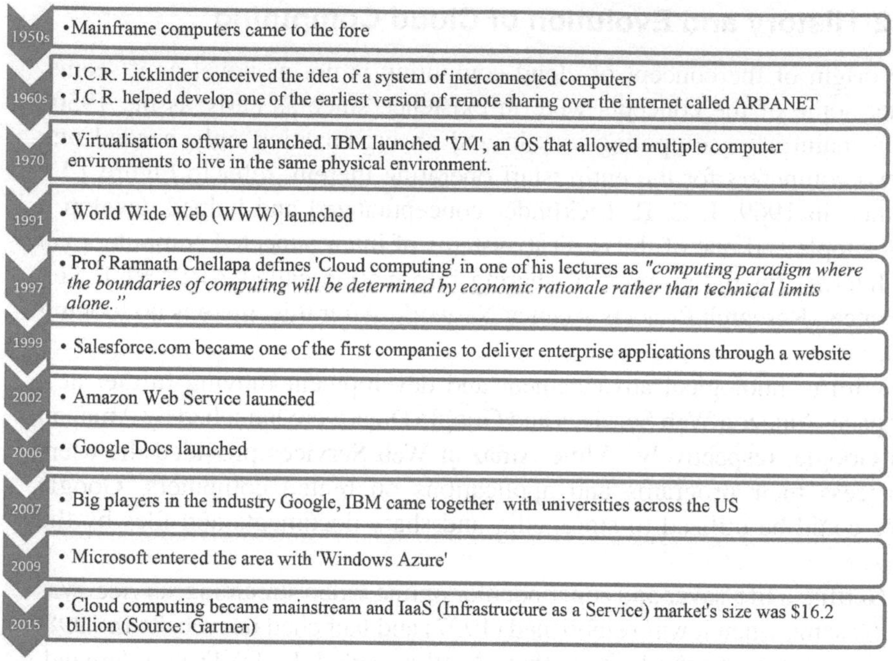

1950s	• Mainframe computers came to the fore
1960s	• J.C.R. Licklinder conceived the idea of a system of interconnected computers • J.C.R. helped develop one of the earliest version of remote sharing over the internet called ARPANET
1970	• Virtualisation software launched. IBM launched 'VM', an OS that allowed multiple computer environments to live in the same physical environment.
1991	• World Wide Web (WWW) launched
1997	• Prof Ramnath Chellapa defines 'Cloud computing' in one of his lectures as *"computing paradigm where the boundaries of computing will be determined by economic rationale rather than technical limits alone."*
1999	• Salesforce.com became one of the first companies to deliver enterprise applications through a website
2002	• Amazon Web Service launched
2006	• Google Docs launched
2007	• Big players in the industry Google, IBM came together with universities across the US
2009	• Microsoft entered the area with 'Windows Azure'
2015	• Cloud computing became mainstream and IaaS (Infrastructure as a Service) market's size was $16.2 billion (Source: Gartner)

Figure 12.1 Timeline of cloud computing.

12.3 Application of Cloud Computing

Multi-pronged application of cloud computing in the public sector:

i. **Executive and legislature:**
   Executive and legislature are more or less responsible to the people and all the bills and acts passed are for the benefit of the people. So, using cloud computing can be beneficial to the executive and legislature in many ways.
   First, the text of the bills can be uploaded on the cloud for anyone from the public or office holder to view easily, irrespective of their geographical location.
   The members of the legislature are the representatives of the masses; often, we frown over the Acts that have been passed by the Parliament with a majority of "members present and voting," but with the help of cloud computing, Parliaments can secure a majority of "all members" as the members who absent themselves on any behest can also cast their votes live in favor of or against a Bill with the help of cloud computing.

Sessions of the Legislature could also be held completely online.

After witnessing a huge number of fatal viruses and diseases that have no cure over the last decade, Governments all over the world have gleaned how important a backup plan is! Amidst the latest pandemic COVID-19, physical distancing has become the new norm, simultaneously being the only known way of prevention. Cloud computing could come in very handy and be an efficient way in which work of the executive and the legislature could continue.

Cloud computing could also be a cost-effective and proficient way of increasing the public involvement in law making by setting up a "Public Feedback Mechanism" so as to ensure public support and promote transparency in the working of the Governments.

ii. Judiciary (courts):

Recently, in the wake of COVID-19, the courts have been striving hard to minimize physical contact and, thus, have switched to video conferencing as well as e-filing. Cloud computing will be a boon for e-filing. It will not only reduce infrastructure costs and be an environmentally sustainable practice but will also help in proper maintenance of records with the Courts. Further, the Counsels or the clients can anytime view the documents that form the part of the record or evidence of their cases, by logging into the cloud. The records will also be safe from all natural or man-made disasters or accidents.

In addition, in cases of appeal, the Court will simply have to transfer the file to the higher Court that has all the records, evidence, and documents present therein, thereby ensuring credibility and transparency.

Further, if all agencies enter document registration details and save it in the cloud, the documents being submitted in the Courts can easily be verified, thus reducing the chances of forged documents being submitted in the Court.

In developing countries, where corruption persists and work is still being carried out in the tradition pen-paper way, cloud computing can ensure that the evidence/testimony is being saved on the cloud; thus, it cannot physically be destroyed and an airtight case can be built around the accused.

In addition, the officers of the Court, i.e., the lawyers can also "be present in the Court" virtually, from home through video conferencing, without the hassle of going to the Court every day saving infrastructure costs of chambers and more staff requirements at hand.

All the virtual hearings can be recorded and a digital library can be maintained, by using cloud computing, for each case ensuring transparency and accountability, with the exception of cases pertaining to sensitive issues and/or national security. The recording will not only serve the purpose of transparency but also be highly beneficial to budding lawyers to watch and learn from. In various countries, such practices have already been adopted. For instance, in Australia,[4] judicial proceedings are recorded and posted on the website of the High court, while in Canada[5] and England,[6] the proceedings being held at the Supreme Court can be live streamed.[7]

The live stream and/or availability of recording of court hearings will also be extremely beneficial in ensuring that due process of law is followed, justice is done, and no bias is exercised in prosecuting persons involved in cross-border disputes wherein a wrong has been done under the territorial jurisdiction of another country or citizen of another country.

It would be pertinent to mention the 2012 Enrica Lexie case,[8] wherein two Italian marines onboard a commercial Italian oil tanker killed two fishermen off the coast of Kerala, India. The two Italian marines were detained in India for a few years after which they were sent back to Italy. Further, citing functional immunity at the time of the incident, the Permanent Court of Arbitration decided that the trial take place in Italy.[9] In this case, live streaming or recording of the court proceedings can be accessed by officials of India via cloud computing and keep themselves updated as to the status of the judicial proceedings.

Cloud computing can also be used to provide services like translation of the judgments/orders/decrees in local dialect and/or international languages in the spirit of inclusivity and awareness.

---

[4]https://www.supremecourt.vic.gov.au/about-the-court/webcasts-and-podcasts accessed on 09/0820 at 16:30

[5]https://www.scc-csc.ca/case-dossier/info/webcasts-webdiffusions-eng.aspx accessed on 09/08/20 at 17:35

[6]https://www.supremecourt.uk/live/court-01.html accessed on 10/08/20 at 18:16

[7]https://www.livemint.com/Politics/GrWwxGSHTseJAEfinR8AwI/Supreme-Court-allows-for-livestreaming-for-cases-of-constit.html accessed on 10/08/20 at 18:13

[8]https://www.thehindu.com/news/international/india-should-release-italian-marine-un-arbitration-court-rules/article8547444.ece accessed on 08/08/20 18:52

[9]https://pca-cpa.org/en/cases/117/ accessed on 10/08/20 at 18:46

iii. Law enforcement:

Cloud computing can easily be utilized by law enforcement agencies for the maintenance of a database regarding criminals and their crime by a two-dimensional approach; first, the repository can be accessed domestically by all law enforcement departments so as to increase the rate of apprehension of criminals, and, second, the repository will also help in apprehension of criminals. Internationally, with a global database available, better coordination amongst countries, and extradition treaties, more criminals can be brought to justice including the wrong-doers involved in financial frauds/crimes, criminal acts, and acts of terrorism.

At the international level, terrorism is a common enemy that various countries are bedeviled by. The use of cloud computing to create a global database of terrorists which can be accessed by intelligence agencies of countries committed to eliminate this plague will also assist in realizing the objective of the 1999 International Convention for the Suppression of the Financing of Terrorism[10] which seeks to cap terrorism by criminalizing the act of funding terrorist activities. In addition, the database could also contain information like their affiliations, their last known location, last movement recorded, insights of experts useful in their apprehension, etc.

Also, it is quite helpful in solving crimes of human/child trafficking as it is often a cross-border crime, wherein people abducted from one country are trafficked across international borders.

Cloud computing can also be used to establish a direct and prompt line of communication in cases of extradition between various federal and foreign law enforcement agencies involved.

iv. Public:

The public can also benefit greatly with the use of cloud computing in the public sector. However, with the introduction of cloud computation, a few developing countries might face issues as everyone might not be well-versed in technology as developed countries.

But this can be used to their advantage by the developing countries, as this will not only act as motivation to increase the literacy rate but also encourage more and more people from all walks of life, and come forward and learn the technological know-how required for availing services of the public sector.

[10] https://treaties.un.org/pages/ViewDetails.aspx?src=IND&mtdsg_no=XVIII-11&chapter=18&clang=_en accessed on 10/08/20 at 17:50

Data can easily be stored on the cloud and accessed by relevant government agencies whenever needed, which can be used to credit social security benefits right in the beneficiary's bank account. It will also be easier to maintain a record and check previous transactions by both the beneficiary and the Government agency passing on the benefit. The use of cloud computing to pass on benefits, if done systematically, can ensure transparency and ease along with the reduced role of intermediaries, shorter time spent in documentation, accountability, and authentic status of social security and whether it is actually benefitting the beneficiaries or not.

A citizen's registry can also be maintained on the cloud, with full access to government agencies, while limited on-request access to private players. Citizen's registry will maintain a record of all the citizens, including their credentials and past transgressions. This will be helpful in doing background checks on people before appointing them at key government positions. At the same time, citizens can have limited on-request access to know the candidates contesting for elections so as to make informed decisions about them.

The citizen's registry facilitating a background check will also aid in obtaining credit from public sector financial institutions with more ease, in a prompt and transparent manner.

In addition, cloud computing can also be utilized in providing "green shoots" to the employment sector by maintaining a portal with authenticated and verified individuals, i.e., both service seekers and service providers, which can be matched with each other based on their qualifications and requirements.

Recently, the President of the United States, Mr. Donald Trump, announced that the November 2020 general elections in the US may be delayed owing to the threat to people's health and safety amidst these pandemic times.[11] Even though the authority to delay the elections does not lie with him and it is only the Congress that determines the date of election and further by an *1845 federal law the date has been fixed as the 'first Tuesday after the first Monday in November.*[12]

Universal mail-in voting is an option which has been opted for by voters where they, due to any reason, are unable to make it to the polling booth

[11]https://www.nytimes.com/2020/07/30/us/politics/trump-postpone-election.html accessed on 12/08/20 at 19:30

[12]https://www.loc.gov/law/help/statutes-at-large/28th-congress/session-2/c28s2ch1.pdf accessed on 12/08/20 at 20:10

to physically cast their votes. *In the 'Universal mail-in voting' system, the registered voters request an absentee ballot from their election office after which the office mails them the ballot. The voters have to provide their name and address and enclose their vote in the security envelope provided with it which aids in retaining the secrecy of the ballot. Once posted by the voter, on receipt by the election office, the officials check the details of the voter after which they remove the sealed ballot from the outside envelope and the vote is counted along with the other mail ballots on Election Day, and is added to the other votes that were cast physically.*[13]

The universal mail-in voting system is quite a respite for senior citizens and people eager to vote but physically unavailable at the place where they are registered to vote. However, the lacuna of corruption and susceptibility to manipulation and tampering make universal mail-in voting a dubious choice.

This is where cloud computing comes to your rescue. Cloud computing, in a secure cloud and air-tight security with chances of no security breach, can be used by voters to cast their votes in real time and live during the election, irrespective of their territorial jurisdiction. It is a win-win situation for democracies that are eagerly waiting to hold elections but are unable to, owing to strict social and physical distancing norms.

v. **International arena:**

Cloud computing in the public sector at the international level is as beneficial as it is at the domestic level.

At the international level, with a shared virtual space, it will be easier and prompt to establish a line of communication. Also, more opportunities of cooperation, development, and peaceful coexistence may get a chance to thrive.

The shared virtual space can be most beneficial for developing countries as a way to interact, learn, and undertake developmental activities that might benefit their country.

Extradition is an intricate procedure even when a treaty between countries exists; imagine what a task it must be, if a treaty between

[13] https://www.brookings.edu/policy2020/votervital/how-does-vote-by-mail-work-and-does-it-increase-election-fraud/ accessed on 14/08/20 at 20:55

countries does not exist! Cloud computing at the international level can also be used to make the process of extradition less cumbersome and as swift and effective as possible, so as to impart justice.

Crimes like terrorism and human trafficking can also be curbed if the entire international community commits to the cause and undertakes to cooperate with each other without any ulterior motive. Making a repository, surveillance, tipping off, etc., can easily help in apprehending the guilty and saving the lives of many.

Resolving cross border disputes in an amicable way can also be promoted through the use of cloud computing by monitoring the proceedings of the court as explained above in the case of Enrica Lexie or establishing a line of communication via video calls, meetings, etc.

Climate change is an emergency, which can be tackled well by promoting the use of virtual resources rather than physical products. In addition, countries can share their environmentally sustainable technologies, indigenous or otherwise, which can be used by other countries.

12.4 Advantages of Cloud Computing

i. **Reduced costs:**

"*Half of all CIOs and IT leaders surveyed by 'Bitglass' reported cost savings in 2015 as a result of using cloud-based applications.*"[14]

The use of cloud computing can also help the public sector in cutting capital expenditure costs and diverting the saved costs toward citizen welfare and other developmental activities.

Computers are an absolute must nowadays; however, as the storage space on a good quality device increases, so does its price tag. Thus, if an organization opts for cloud computing, it is not only saving on maintenance and upgrading costs of equipment but also large teams of IT experts hired to manage and store the data. Instead of paying an entire team of at least 7–10 people, a medium-sized organization, if operating on the cloud, can often make do with a small yet efficient team of 2–3 professionals, in cases where digital literacy of other employees is low.

[14]https://www.salesforce.com/in/products/platform/best-practices/benefits-of-cloud-computing/ accessed on 16/08/20 at 16:20

Benjamin Franklin said, "Time is money"; with cloud computing, we see lesser energy consumption (due to lesser infrastructural uses) and fewer delays, and one can also work on-the-go from any device of their choice, be it a smart phone, a laptop, or a smart pad. Moreover, through cloud computing one has access to their data and work files, irrespective of wherever they are, provided they have Internet access; hence, it is a huge time saver or a money saver, to be precise.

Most of the subscription plans offered by cloud service providers are based on the "pay-as-you-go" model which ensures that the user pays only what he has signed up for and any extra services that they do not need or want do not need to be paid for.

Hence, the cost–benefit ratio of utilizing cloud computing is higher than the traditional method of storing and managing data in the pen and paper mode or the usual operating systems in offices.

The saved public funds can instead be utilized to fund other requisite developmental activities.

ii. Scalability:

Investopedia defines scalability as the ability to manage well with an increased or decreased workload or scope.[15]

Every organization has different needs; for instance, the needs of an office of finance dealing with taxes, licenses, fees, permits, etc., with a workforce of 500+ employees having jurisdiction over a sprawling metropolitan area, will definitely differ from the Finance Department having a smaller workforce having jurisdiction over a village or a suburban town.

One of the most beneficial features of cloud computing is its scalability which enables it to be super flexible and match up to needs of the organization it is attached to, without incurring any substantial extra costs owing to expensive equipment installation/upgrade. Cloud computing enables you to have high performance resources and professional solutions, on demand.

For example, cloud computing can tackle your demand for extra bandwidth instantaneously, without going through the process of a cumbersome update to your entire IT system.

Crashing of public sector websites during an important task with a short deadline is not completely unheard of and can be remedied with the use of cloud computing.

[15]https://www.investopedia.com/terms/s/scalability.asp accessed on 16/08/20 at 14:53

Another apt example relates to one of your sales promotion being quite popular; capacity can instantaneously be added to it so as to avoid the risk of losing sales and servers crashing. And when the sales are done, the capacity can also be shrunk for the reduction of costs.[16]

iii. Better collaboration and coordination:

Even though the public sector has an office for every diverse set of works it deals with, however, in some cases, two or more offices may be intertwined.

For example, an extradition of a criminal is a very intricate process which involves various law enforcement agencies, both federal and foreign.

When seeking extradition, the United States alone involves various stakeholders and agencies at every stage. *Firstly, the **State or Federal prosecutors** communicate and discuss the charges, etc. with the **relevant law enforcement agency**. The Prosecutors who seek the extradition must get in touch with the Criminal Division's **Office of International Affairs (OIA).***

OIA reviews and approves every official request for international extradition based on federal criminal charges. And OIA also does the same at the request of the State Department in case of requests for international extradition based on State charges.[17]

*Once they decide to go forward with it, the Prosecutor's Office submits an application to the **Justice Department** for review and examination. Once satisfied, the Justice Department forwards it to the **State Department**, which then sends it to the relevant **U.S. Embassy** who is tasked with sending the application to the **relevant authorities in the country** from where US seeks extradition of the person accused.*

*On the other hand, when the U.S. receives a request for extradition, **the foreign government** sends in a request to the **U.S. State Department** with relevant documents including treaty information, details of the person to be extradited, charges, evidence, arrest warrant, etc. If **the Secretary of State** deems fit, the request is forwarded to the **Justice Department**, and is examined and contested before a **federal judge or a magistrate**. It isn't a proper hearing or trial therefore the rights of*

---

[16]https://www.idexcel.com/blog/top-10-advantages-of-cloud-computing/ accessed on 12/08/20 at 17:00

[17]https://www.justice.gov/jm/jm-9-15000-international-extradition-and-related-matters accessed on 16/08/20 at 18:09

the person accused are limited, it is a hearing only to determine whether there is probable cause that the person has committed the alleged offence or not. Once the court is satisfied that the probable cause is present, it certifies extradition and sends the certification to the Secretary of State, who takes the final call on the issue.[18]

Extradition is indeed an interwoven and convoluted process, which can be made easier, cost-efficient, and prompt by the use of cloud computing for accessing files, establishing a direct line of communication between various stakeholders involved, even while sitting miles away from each other, keeping a check on the process, etc.

iv. **Users can work remotely:**

Even though nobody likes to work on a vacation or a public holiday, yet exigencies in public offices can arise any time. So cloud computing deals with such exigencies fairly well as it provides access to your virtual office, i.e., the files and data, irrespective of your geographical location. The only pre-requisite for the same is an Internet connection, and you can access your work anywhere, anytime, and on any device, be it a smart phone or any other device.

It is extremely beneficial in cases where:

a. employees have substantial field work;
b. where freelancers have been employed;
c. while outsourcing the work;
d. while employees are on leave;
e. during any emergency situation coinciding with a public holiday; or
f. while working from home.

v. **Data security:**

The phrase – "The show must go on" even though is popularly used for show business – equally applies to any organization that provides people services nowadays, especially public sector organizations.

During any disaster, it is these public sector organizations in whom the people repose their trust, to be well taken care of and be helped in getting through the same. Thus, even in cases of a disaster, natural or man-made, the work (show) of these public sector organizations must (and actually, does) go on!

[18] https://www.cfr.org/backgrounder/what-extradition accessed on 15/08/20 at 18:01

Data loss in this age is a huge setback to any organization as pretty much everything including the names of beneficiaries, their details, etc., is stored in the form of data on work computers. If any cataclysmic disaster, be it natural or man-made, any power outage, or any other catastrophe damages the "work computers," the data is forever lost. This is where cloud computing comes in handy as even if the work computers are damaged, the data is saved in the cloud and can be accessed and retrieved from any other functional computer or device with Internet access. The services provided by various cloud-based service providers are impeccable as they maintain an uptime almost 99.9% of the time, 24 $\times$ 7, and some applications even function offline.

In addition to loss prevention, cloud-based service providers assist and enable users in prompt data recovery, irrespective of your location and device.

Sensitive information or confidential files which should not end up in the wrong hands can also be saved in a private cloud with limited access, instead of saving it in the work computers, eliminating the risk of losing it or it ending up in the wrong hands.

Also, *cloud storage providers implement baseline protections for their platforms and the data they process, such as authentication, access control, and encryption.*[19]

vi. **Easy management and better control:**

Cloud-based service providers take away an organization's headache with respect to management, maintenance, and upgrading the IT infrastructure. Cloud gives you a basic and simple user interface without the need for installing it, thus ensuring easy management.

Further, an organization has better control over their reporting as all the files have the same format and are available at a central storage location with access to all relevant employees; thus, human errors, inconsistency in data, double reporting, etc., can be ruled out to a huge extent.

In addition, the person in charge has complete control over the data including aspects like which files are accessible by all, which files are accessible by only limited employees or on a need to know basis. This

[19]https://www.globaldots.com/blog/cloud-computing-benefits accessed on 14/08/20 at 18:57

guarantees better security and also cuts out the work in a well-organized way for employees.

vii. Environmentally sustainable:

Most of us acknowledge the fact that the environment is in danger, but are we doing enough to save it?

Shifting from physical products to virtual services is an efficient way to reduce the carbon footprint of an organization. Shifting to cloud computing helps the environment by reducing energy consumption; using only the resources we require; reducing e-waste; vehicular emissions; reducing paper usage, amongst others.

"*A Pike Research report predicted data centre energy consumption will drop by 31% from 2010 to 2020 based on the adoption of cloud computing and other virtual data options.*"[20]

12.5 Challenges

i. Security breaches:

"*As more and more businesses migrate more and more of their services and functions to the cloud, security and regulation compliance will become an increasing concern.*"[21]

Recently, on July 15, 2020, approximately 130 high profile Twitter accounts (including Joe Biden, Elon Musk, Barack Obama, etc.) were hacked by a team of hackers, who then attempted (and were, to a certain extent, successful) a bitcoin scam which was created to steal money from regular Americans. It was a big security breach and it took Twitter hours to get a hold of their site, and restoring some of the compromised accounts took days.

Even though we are told, time and again, to be cautious when we put personal data, social security details, bank details, etc., on online websites, but one lucrative deal and we inadvertently, without inspecting and verifying the deal and/or website, jump right in.

Data is money, in this time and age. Data is stolen and often sold on the "dark web."

[20]https://www.salesforce.com/in/products/platform/best-practices/benefits-of-cloud-computing/ accessed on 16/08/20 at 19:15

[21]https://the-report.cloud/the-evolution-of-cloud-computing-wheres-it-going-next accessed on 16/08/20 at 11:15

If a citizen's (who has nothing to do with national security) data gets compromised and there is a massive market for that, consider how big of a market there must exist for data, that directly or indirectly, relates with the national security of a country.

Public sector organizations owing to the functions performed by them are a huge repository of citizens and sensitive/confidential information related to them. Specialized public sector organizations also deal with the national security of a country. If everything is shifted to the cloud, we have a potent remedy for data loss due to any natural or man-made exigencies, but what remedy do we have against data theft or a possible security breach?

To prevent security breaches in the cloud, extra security measures will have to be taken by the organizations; otherwise, they are simply a sitting duck.

According to the 2013 CDW State of the Cloud Report, *46% respondents cited security of data or applications as a prominent challenge.*[22]

At times, organizations fail to exercise due diligence and do not enquire about the security measures being taken by the cloud or the scope of protection being given to them and their data. Mostly, cloud-based service providers are responsible for the security of the cloud but not for the apps, servers, and data put out there by you.

To bolster data protection, an extra layer of encryption and securing the data needs to be added.

ii. Lacking a uniform regulatory framework:

Most of the cloud-based service providers base their data centers at affordable geographical locations; however, they cater to people across the globe with the services they provide. Different countries have different laws that govern the residents of that country and the data of those residents stored by the cloud-based service providers. For instance, some laws of the United Kingdom specify that the data related to UK citizen can only be stored within the country.

Another instance comes from the US. Due to the implications of the U.S. Patriot Act, government agencies can easily access the data of a

[22]https://cdw-prod.adobecqms.net/content/dam/cdw/on-domain-cdw/cdw-branded/newsroom/tech-insights/February-2013-CDW-2013-State-of-The-Cloud-Report.pdf accessed on 16/08/20 at 11:00

European citizen stored on the cloud in the US.[23] But the same is not the case if the data is stored in few of the European countries.

Further with effect from May 2018, the European Union's legislation on data protection called General Data Protection Regulation (GDPR) was enacted. The purview of GDPR is wide as it takes into consideration not only those companies that are based in the European Union but also companies that deal with an EU citizen's data, irrespective of where they are based.[24]

While CCTV cameras are the way of life in some countries, but it is also collection of personal data on a hard drive or on the cloud. And even before the GDPR was enacted in EU, the installation of CCTV cameras is regarded as infringement of privacy.

In the case of **TK v. Asociaţia de Proprietari bloc M5A-ScaraA**[25], which was referred to the CJEU by the Romanian Court for guidance, residents of an apartment building objected to the installation of CCTV cameras in the common areas of the building owing to infringement of the right to privacy. The question that arose before the Court was that whether in light of *Articles 6(1)(c) and 7(f) of the Data Protection Directive (95/46/EC), read along with Article 7 and 8 of the EU Charter of Fundamental Rights*, the national law allows the installing of a video surveillance system in a residential building on grounds of "legitimate interests" (which, in this case, were the safety and security of the residents and their property) without their consent?

The CJEU in the case observed that the continuous recording as done in video surveillance system amounts to automatic processing which should, first of all, adhere to the principles mentioned under Articles 6 and 7 regarding data quality and making the processing legitimate, respectively.

The more important part here relates to the "legitimate interest" basis which, in this case, is the security of residents and their property as

[23]http://whatiscloud.com/cloud_deployment_models/ accessed on 17/08/20 at 10:00

[24]https://www.geospatialworld.net/blogs/check-out-10-key-features-of-gdpr/#:~:text=The%20GDPR%20is%20all%20set,power%20over%20their%20personal%20data.&text=The%20new%20law%20will%20apply,those%20dealing%20with%20EU%20citizens accessed on 17/08/20 at 10:15

[25]EU:C:2019:1064 (C-708/18); http://curia.europa.eu/juris/document/document.jsf?text=&docid=221465&pageIndex=0&doclang=en&mode=lst&dir=&occ=first&part=1&cid=7340728 accessed on 18/08/20 at 18:20

numerous incidents of breaking and entering and vandalism had been reported.

The CJEU further mentions the "purpose test," i.e., the data controller or any party the data is being disclosed to must be acting in lawful pursuit of a legitimate interest; the "necessity test," i.e., there must be an absolute need to process personal data owing to the legitimate interest being pursued; and the "balancing test," i.e., the Fundamental rights of the individuals concerned do not override the legitimate interest. Further, it is also observed that the legitimate interest being alleged should be present at the time of processing and the data minimization principle mentioned under Article 6(1)(c) should also be borne in mind.

The CCTV camera footage can be stored on any cloud-based provider, however, differing levels of stringency in laws and regulatory provisions in different countries, are a hassle for all stakeholders involved.

In the Google Spain case, i.e., **Google Spain SL, Google Inc. v. Agencia Española de Protección de Datos (AEPD), Mario Costeja González,**[26] the "right to be forgotten" or "the right to erasure" had been mentioned for the first time.

Later, in the 2016 case of **Google LLC v. Commission nationale de l'informatique et des libertés (CNIL),**[27] the "right to be forgotten" was mentioned again along with an elaborate explanation on how it can be effected. In the present case; the CJEU stated that the territorial application of right to erasure or be forgotten will only be limited to the Member States of the European Union, and when being exercised against a corporate with multi-national operations, it can only be effectuated with respect to operations that have been undertaken in the EU. The case was referred by the French Council to the ECJ and pertains to Google's refusal to pay the fine imposed on it in 2016 by the French data protection regulation, CNIL. Google's refusal to CNIL's notice asking them to de-reference (remove) links that contained personal data of individual citizens, on its search engine worldwide, was the reason for the imposition of the Ă100,000 fine on it.

---

[26]EU:C:2014:317 (C-131/12); http://curia.europa.eu/juris/document/document.jsf?text=&docid=152065&pageIndex=0&doclang=en&mode=lst&dir=&occ=first&part=1&cid=15404758 accessed on 20/08/20 at 11:11

[27]EU:C:2019:772 (C-507/17); http://curia.europa.eu/juris/document/document.jsf?text=&docid=218105&pageIndex=0&doclang=en&mode=lst&dir=&occ=first&part=1&cid=15296989 accessed on 20/08/20 at 19:30

The ECJ referred to *Article 17 of the General Data Protection Regulation (GDPR) along with Article Articles 12 and 14 of the Data Protection Directive, and the seminal 2014 Google Spain judgment, which first recognized the "right to erasure" under existing EU law.*

Many states do not acknowledge de-referencing or might have a completely different perspective on the issue, was also taken into account by the ECJ and it held that the de-referencing by search engines, Google or otherwise, will be done from the domain within the EU and not worldwide.

It is pertinent to mention here that GDPR is such an immaculate legislation that it even provides for data deletion or the right of de-referencing, in case the individual withdraws consent or if the information does not hold any truth or relevance; but as mentioned in the judgment by the ECJ, many states are not aware or do not provide such stringent data protection regulations to their citizens.

On the other hand, the People's Republic of China (PRC) initially lacked a single data-centric legislation, but data protection issues were dealt with, under the provisions of civil law and tort law as matters of either reputation or privacy. However, in 2017, it enacted the PRC Cyber-Security Law. Under the PRC Cyber-Security law, consent of the data subject for collection of data has been emphasized upon along with its security. Further, the law also prohibits the illegal collection, processing, usage, transmission, buying/selling, or disclosure of other person's personal data. The scope of application of PRC's cyber security law is similar to the GDPR in respect of the fact that companies with multi-national operations will comply with the PRC's laws and regulations while dealing with residents of PRC, irrespective of where such companies are based. Companies collecting or processing data in addition to consent will also have to explicitly mention their purpose for collection of data and only the data collection necessary for the purpose for which consent has been given must be collected.

The law also provides for data deletion and specifies therein that the data collected can be stored for the minimum duration in pursuance of the purpose and thereafter the data so collected and stored will be deleted properly, which is synonymous with GDPR's right to be forgotten.

Similar to the UK, PRC's cyber security law also requires that the personal data collected and generated in China by Critical Information Infrastructure (CII) operators be stored within the territorial limits of the PRC. However, it allows conditional cross border transfer of data by CII

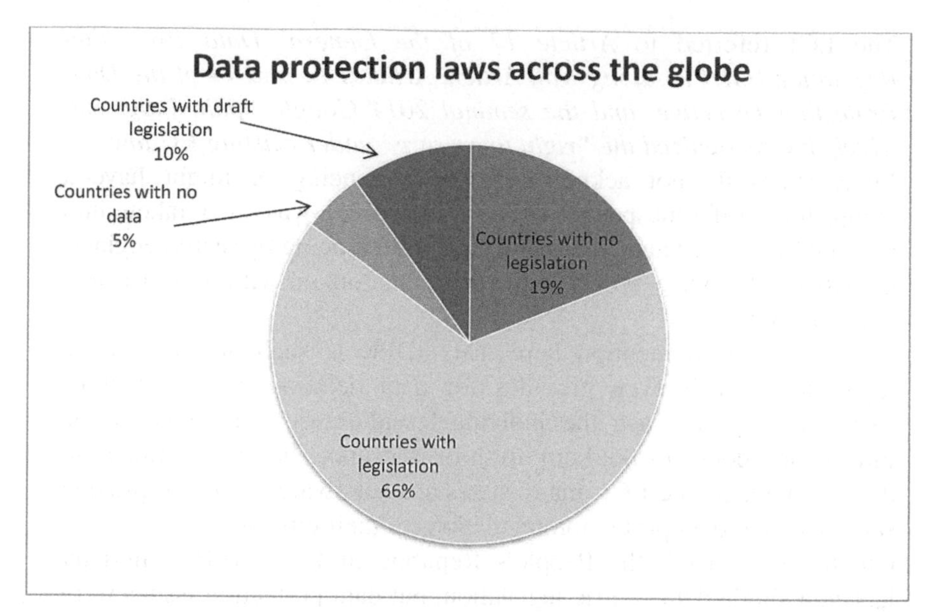

Figure 12.2 Data protection laws around the globe

category network operators if absolutely necessary for the purposes of business.
Different countries and different regulations definitely do not make the use of cloud computing a hassle-free task.
Thus, uniform regulatory framework should be provided for cloud-based service providers and data protection so that not just EU citizens are protected but everyone around the globe is.

iii. Better laws on data protection:
Figure 12.2 above represents the current state of data protection laws according to UNCTAD.[28] 29% of the countries even in 2020 do not have legislation on data protection (draft + no legislation).
Unless every country has a strict policy and law to deal with protection of data, data will get stolen and leave us in a vulnerable position.
As mentioned above, various developed counties like U.K., Australia, the EU Member States, etc., have stringent data protection regulations for their citizens owing to the fact that any leak of personal data is detrimental to an individual. Such developed States have prescribed

[28]https://unctad.org/en/Pages/DTL/STI_and_ICTs/ICT4D-Legislation/eCom-Data-Protection-Laws.aspx accessed on 17/08/20 at 08:00

for the subject's consent, right to be forgotten, stringent measures for storage of data, highly regulated framework for cross-border data sharing, etc.

So does it imply that only the citizens of developing countries are left out in the cold to be exploited?

iv. **Lack of digital literacy:**

In developing countries, the rate of digital literacy is low. Thus, if the public sector integrates cloud computing, it might be beneficial only to the percentage of the population that is digitally well-versed. For the large percentage of the population who are not well-versed in how things work online, it will be a difficult task for them to obtain any benefit from the new system.

In addition, integration of cloud computing in the public sector organizations will need skilled staff having a certain level of technical know-how as to how things work. Training can help the employees who have basic knowledge; however, employees with no knowledge will have a hard time coping with the system, in turn, affecting their job satisfaction and efficiency.

Also, if the employees find it a hard task to conduct simple tasks via the cloud, a team of expert IT professionals will have to be employed, which will defeat one of the major benefits of cloud computing, i.e., reduced IT costs.

Lack of digital literacy is a major barrier in moving toward cloud computing in an organization; however, it can also be treated as a motivating factor boosting the rate of literacy as well as digital literacy in a country.

Amongst other major challenges are vendor lock in, poor Internet connectivity, cyber security risks, unauthorized access by employees, risk of losing sensitive/confidential stored on the cloud, etc.

12.6 Conclusion

Cloud computing has come a long way. Gartner Forecasts predicts the revenue forecasts to rise from $206.2 billion in 2019 to $240.3 billion in 2020 and $278.3 billion in 2021.[29]

---

[29]https://www.gartner.com/en/newsroom/press-releases/2018-09-12-gartner-forecasts-worldwide-public-cloud-revenue-to-grow-17-percent-in-2019?ref=hackernoon.com accessed on 17/08/20 at 09:00

However, egregiously, we still live in a world where 5% of the countries have no data. Countries are still segregated into developed and developing countries.

Undoubtedly, cloud computing is the behemoth taking over the market, but all good things come at a price.

As we see an increase in individuals and institutions shifting to the cloud, risks also seem to be increasing, after all hackers go where the data goes.

Integration of cloud computing in public sector organizations and its extent should be determined after taking all substantial factors into account including population's needs and their technical know-how, technical know-how of employees, cost of training the staff, safety considerations, type of data intended to be stored in the cloud, etc.

Usage of cloud computing will not only be beneficial at the domestic level but also at the international level as discussed earlier; however, it should be rolled out in a phased manner. In addition, public sector organizations, when it comes to data security and protection, need to exercise due diligence and strictly abide by the doctrine of "caveat emptor."

13

Big Data Analytics: An overview

Dipalika Das[1] **and Maya Nayak**[2]

[1]Department of MCA, Trident Academy of Creative Technology, BBSR, Odisha, India
[2]Department of CSE, Orissa Engineering College, BBSR, Odisha, India
E-mail: e-mail:dipalika.das@gmail.com, mayanayak3299@yahoo.com

Abstract

Today, we are in an information era where large amount of data are available for decision making. Those datasets which are huge in volume, fast in velocity, and large in variety are known as big data. Existing tools and technologies are not sufficient enough to handle such a big data and, hence, would face difficulty. As data is increasing with rapid pace, several new methods of solutions are studied, analyzed, and implemented to obtain essential hidden values and knowledge from the dataset. Analysts and researchers work hard to get valuable information from such huge sources of data (social network data, customer interactions with business sites, etc.). Big data analytics is nothing but the application of advanced analytic techniques on big data. This paper is a step toward discovering different analytic tools and techniques that would be beneficial while dealing with big data in various fields.

Keywords: Big data, analytic method, knowledge, datasets, transaction.

13.1 Introduction

As there is increasing demand for Internet, Internet of Things (IoT), and cloud computing, it leads to exponential growth of data in almost every sector like industry, business, etc. every year. The innovative use of technologies and creativity to provide the right data from an exponentially growing storage to the right user at the right time is known as big data. Now it is a challenge to deal with the rapidly increasing voluminous, heterogeneous, complex, and interconnected data. This paper puts light on concept of big data, its importance, applications, and challenges in today's scenario.

13.2 Related Work

Huangke Chen *et al.* [1] derived two theorems in order to eradicate the difficulty in workflow scheduling problem by minimizing the completion time and start time of workflow tasks there by reducing the cost. They proposed a new real-time scheduling algorithm using task duplication (RTSATD) based on these two theorems, such that it minimized both the completion time and the cost of processing big data workflows in clouds. The performance of this algorithm was analyzed. The experimental results proved the superiority of the proposed algorithm with respect to completion time and resource utilization.

Mohammad Sultan Mahmud *et al.* [2] presented several partitioning and sampling methods to perform data analysis on high availability distributed object oriented platform (Hadoop). They started using frameworks of big data. They used three data partitioning methods along with three horizontal partitioning techniques:

- Range
- Hash
- Random

Then data sampling methods were verified using:

- Simple random sampling
- Stratified sampling
- Reservoir sampling

They discussed regarding record-level sampling and block-level sampling and found that performance of block level sampling was better than record-level sampling. After analyzing block-level sampling, they could find that it was not suitable for approximate computing. Finally, they concluded that partitioning and sampling on data should be done together to construct framework for approximate cluster computing.

Milad Makkie *et al.* [6] worked to solve the challenges of neuroimaging in big data. They created a data management system to organize the large-scale fMRI datasets and presented their new methods for the distributed fMRI data processing that employed Hadoop and Spark. Finally, they demonstrated the significant performance gains of their algorithms to perform distributed dictionary learning.

Ming Jian Tang *et al.* [7] proposed a new framework. They started tracing out how to handle persistent volatility in the data. They could find out multivariate dependency structure amongst different vulnerability risks. Using the real vulnerability data, they had shown that a composite model could effectively capture and preserve long-term dependency between different vulnerability and exploit disclosures. In addition, the paper provided a way for future study on the stochastic perspective of vulnerability proliferation toward building more accurate measures for better cyber risk management.

Sunil Kumar *et al.* [8] proposed a new clustering algorithm called hybrid clustering in order to overcome the disadvantages of existing clustering algorithms. They compared the new hybrid algorithm with existing algorithms on the basis of precision, recall, F-measure, execution time, and accuracy of results. From the experimental results, it was made clear that the proposed hybrid clustering algorithm was more accurate and provided better results with respect to precision, recall, and F-measure values. Here they had used MapReduce under Hadoop.

Balsam Alkouz *et al.* [9] worked upon a system that would collect raw tweets from tweeter site first. Those raw tweets were processed as follows:

1. Filtration to retrieve influenza related tweets
2. Categorization based on language
3. Tokenization into words

Each processed tweet was categorized into particular class labels like self/non-self-reporting, and non-reporting. By going through the cases related to reporting, the growth of the disease could be visualized. A regression model that was linear could be applied on tweets, related to reporting cases, in order to provide the disease status. It was found that the tweets related to the disease and visits to hospital were highly correlated.

Hao Zhang *et al.* [11] proposed a new system named Generic Manufacturing Data Analytics System (GMDA). Most of the manufacturing data analytics jobs could be easily handled by this new system. The users who had no prior exposure to this area could perform the analysis task efficiently. The new system had its own language called generic manufacturing data analysis task-describing language (GMDL) to properly handle the analysis task required during manufacturing. Their work supported small/medium-scale manufacturers to carry out analysis tasks with their own data and get benefitted even if they did not have prior knowledge regarding data analytics. For the purpose of manufacturing, different algorithms were selected, tested, and then implemented. This system included several important methodologies like:

- Proper algorithm selection strategy
- Optimal parameter determination algorithm

They had taken several example cases to prove the feasibility and efficiency of the system. Their future work was based on RHadoop or SparkR in GMDA to be useful with big data.

Chunqiang Hu *et al.* [12] proposed one access control scheme for big data in cloud. It was found to be secure and based on NTRU cryptosystem. Initially, they used a decryption algorithm based on NTRU. After its successful implementation, they explained its procedure and analyzed its correctness, security feature, and efficiency. The cloud server was able to update the encrypted text as was specified by the data owner. This scheme verified the authenticity of the user before accessing data. It also performed the validation check of the information provided by users. It was found that this scheme could protect users from various attacks and from being cheated.

Xuedi Qin *et al.* [10] developed a new data visualization system called DEEPEYE for visualizing a given dataset. It helped in solving several problems like the following.

Visualization verification: There could be several visualizations possible for selected columns like: bar chart, pie chart, line chart, etc. It was decided whether a chart is meaningful for the selected column or not. Among all visualizations, it helped in finding out which one would be the best. It helped in ranking the good visualizations.

Visualization search space: Data was transformed by using any of techniques like selection, join, binning, aggregation, etc. Once transformation was over, required parts could be selected. This helped in determining the search space.

On-time response: There was provision for on-time visualization. Solving the problem on time was found to be quite expensive because of huge search space and dataset size. For that, efficient algorithms like database optimization techniques, pruning, etc. were devised.

Jun Wu et al. [13] proposed a mechanism called security situational awareness based on big data analytics in the smart grid. In order to carry out security situational analysis on smart grid, they combined together several approaches which included reinforcement learning, game theory, fuzzy cluster based analysis, etc. They used simulations and experiments. Finally, their result showed high performance and less error for the mentioned approach.

Prayla Shyry et al. [15] proposed a cryptosystem named NTRU which enabled the cloud server to refresh the cipher text. It strengthens the information proprietor and qualified clients to confirm the authenticity for getting required information. It helps in approving the data given by different clients to revise plaintext recuperation.

Youssra Riahi *et al.* [14] defined the concept of big data, data analytics, etc. They tried to find out importance, applications, and challenges while learning big data.

Ravi Vatrapu *et al.* [16] proposed a new approach to big data analytics called social set analytics based on social sites like Facebook, Twitter, etc. Social set analysis generally consisted of a generative framework and an analytical framework. Generative framework was meant for the philosophies of computational social science, theory of social data, and conceptual and formal models of social data. The analytical framework was meant for associating big social datasets with organizational and societal datasets. They were able to explain social set analysis in terms of fuzzy set theoretical sentiment analysis, crisp set theoretical interaction analysis, and event studies oriented set theoretical visualizations.

Bharadwaja Kumar *et al.* [17] conducted a detailed study on big data and could list down its various challenges and scope. They had pointed elaborately various characteristics of big data along with solutions to face the challenges that would come on the way. They had used several algorithms of machine learning and data mining to search for the patterns that were hidden in datasets. Finally, they showed the way as to how big data analytics could be useful.

13.2.1 Big Data: What Is It?

The concept behind big data is based on new evolutions and technologies. It helps in providing the right user with right information at the right time from a huge mass of data that has been increasing exponentially. It is a complex polymorphic object. It is designed to provide real-time access to everyone. It is a concept which is difficult to define as the volume of data varies from place to place. This is an emerging field which defines a large number of techniques and technologies.

13.2.1.1 Characteristics of Big Data

Generally, big data is large in size, includes structured, semi-structured, and un-structured data, is more diversified, and is fast in arrival. It can be characterized by two properties:

- 3V
- 5V

3V:

Volume: The system deals with huge data, performs operation on them, and then stores in an organized manner. Volume is directly proportional to the quantity of data collected and stored.

Variety: It shows multiples of types of data handled by the system. There exists a complex link among data collected. It depicts several possible uses in collaboration with raw data.

Velocity: It can be explained as the frequency with which data is generated, captured, and shared.

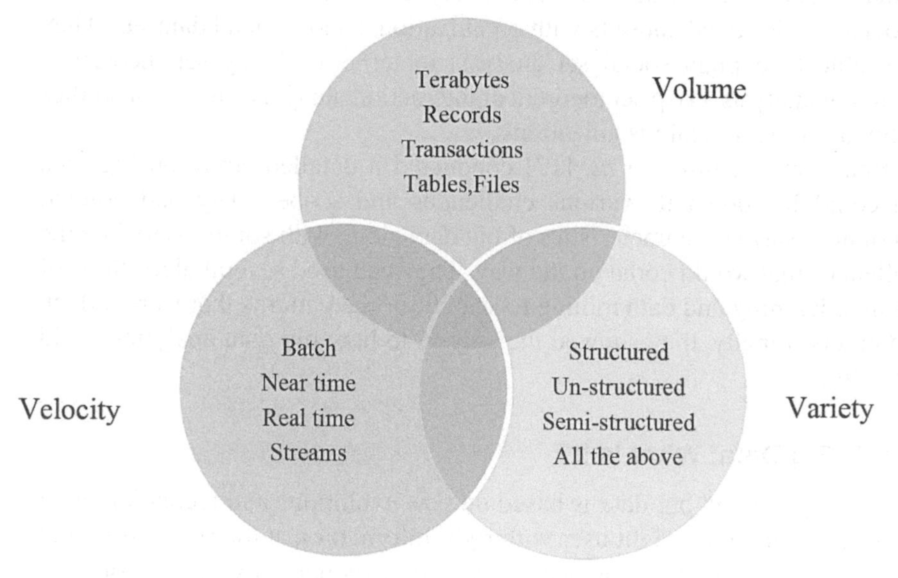

Figure 13.1 3V concept.

5V:

Besides 3Vs discussed above, this feature contains two more Vs:

Veracity: It defines the extent of quality, correctness, and reliability of data and data sources.

Value: It is derived from data that is useful.

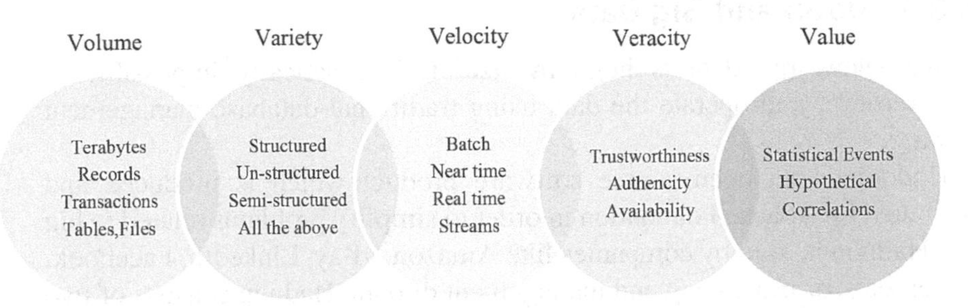

Figure 13.2 5V concept.

13.2.2 Big Data Analytics: What Is It?

It is the process of collection, organization, and analysis of large datasets to search for different hidden patterns and other useful information. It is nothing but collection of technologies and methods to find out hidden values from large datasets. It helps in solving new or old problems effectively and efficiently.

Types:

i. ***Descriptive analytics:*** It starts with a query like: what is happening? It gives an idea regarding what will happen in future. Generally, data mining methods help in organizing the data and finding out useful patterns.

ii. ***Diagnostic analytics:*** It starts with a query like: why it happened? It looks for the main cause of the problem. It determines why something had happened. It tries to identify the cause of events and determine its behavior.

iii. ***Predictive analytics:*** It starts with a query like: what is the most likely to happen next? It prefers previous data to tell about the next possibility of occurrence. This is similar to forecasting. Several methodologies of data mining and artificial intelligence can be implemented to analyze existing data to predict what might happen next.

iv. ***Prescriptive analytics:*** It starts with a query like: what should be done? It is concerned about taking the right step at the right time. It helps in finding the best solution by using information from other analytic methods.

13.3 Hadoop and Big Data

As we know, big data is huge in size, it is practically impossible to keep, retrieve, and operate the data using traditional database management software.

Hadoop is an open source software product which is produced and distributed by Apache Foundation in order to simplify problems related to big data. Hadoop is used by companies like Amazon, eBay, LinkedIn, Facebook, Twitter, etc. for processing and management of data. Hadoop consists of two host servers.

Job tracker: It is one per each cluster associated with Hadoop. MapReduce is run to organize tasks in the cluster.

Task tracker: It is more than one per Hadoop cluster. It executes the map reduce task itself. It reports its status continuously through heartbeat package. If it fails, the job tracker informs the redistribution of task to another node.

It consists of several components like:

- HDFS
- YARN
- MapReduce

HDFS: This is otherwise called Hadoop Distributed File System. This is well aware about the input and output to the MapReduce program. This comprises the following:

1. Name node: It is distinct for a cluster. It stores information about files. This is the main node under which several subordinate nodes exist.
2. Alternate name node: This is vigilant about what is happening in the cluster and keeps track of data in name node. In case the name node fails, the secondary name node takes its place.
3. Data node: This is more than one per cluster. It stores contents of files which are fragmented into blocks.

MapReduce transfers the task to other nodes within the map/cluster. When a query is raised, the outputs from each node are minimized, assembled, and ordered into a single cohesive response. It consists of two functions:

- Map
- Reduce

The data is exchanged between them in the form of pairs (key and value). In response to a pair of inputs to map subroutine, several output pairs are generated. Reduce function accepts a list of input pairs but produces one output most of the time.

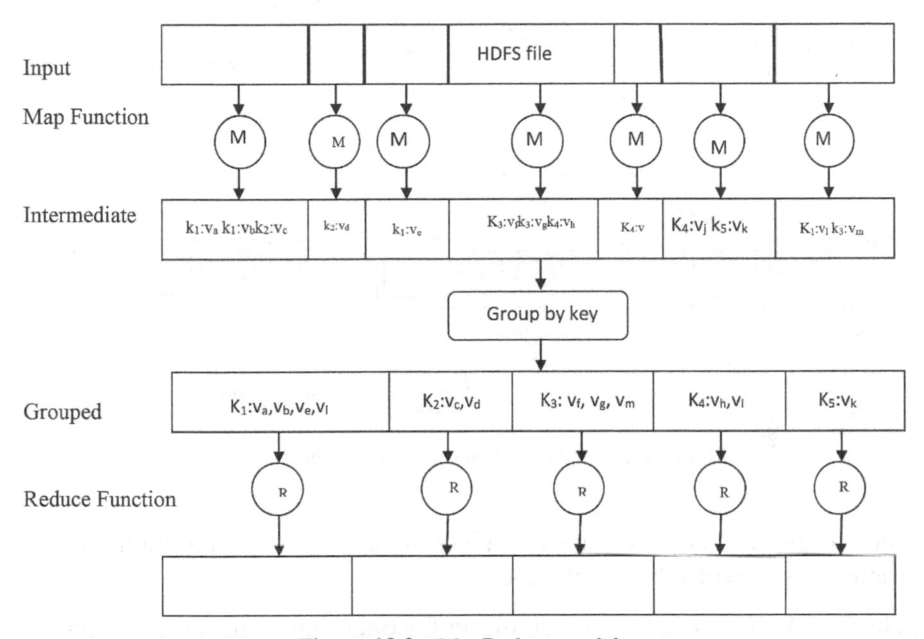

Figure 13.3 MapReduce model.

YARN (Yet Another Resource Negotiator) is located on top of the HDFS to provide operating system capabilities. It manages jobs on a cluster of machines. It monitors progress of each job. It can transfer a job/process from one machine to another.

13.4 Big Data Analytics Framework

Apache Hadoop: This is a platform where processing related to distributed and parallel functionalities on huge data are performed. General partitioning mechanism is applicable on this huge data. Here task is uniformly distributed across several terminals using MapReduce computing methodology. The components of Hadoop platform consists of Kernel, MapReduce, HDFS, YARN (Resource Manager), and several projects like Hive, HBase, etc.

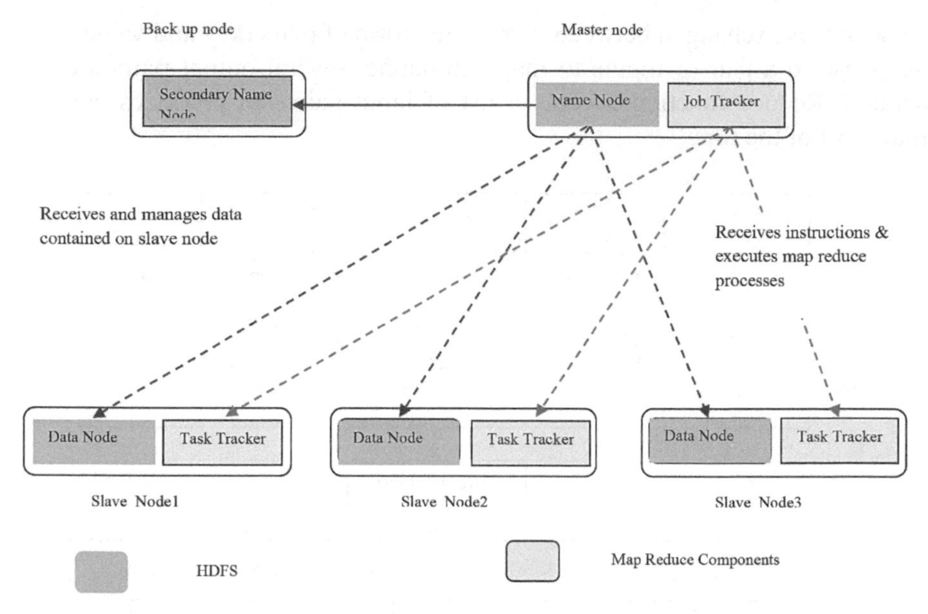

Figure 13.4 HADOOP general architecture.

In order to process large volumes of distributed datasets, a programming environment is provided by MapReduce.

Apache Spark: This is a framework meant for open source data processing. The dataset used is resilient and distributed in nature. Mostly, data is read only, unchangeable, and fault resistant. This consists of various attributes that are distributed across several nodes in a cluster. This supports operations like: transformation (map(), filter()); action (reduce(), count(), collect()).

13.5 Big Data Analytics Techniques

The most commonly used techniques on Hadoop clusters are:

i. Partitioning on big data
ii. Sampling on big data
iii. Sampling-based approximation

13.5.1 Partitioning on Big Data

This is a pre-processing phase which prepares the data for the next analytic phase. When a file containing big data is imported into Hadoop, it is subdivided into tiny fixed size blocks. Initially, HDFS blocks are imported into RDD in Apache Spark which can be partitioned and repartitioned, adopting any of the techniques from hash, range, and custom partitioning. Random sample partitioning (RSP) is a distributed data model consisting of RSP blocks which are non-overlapping in nature. Each of these blocks is nothing but random samples of big dataset. Here two-stage data partitioning method is implemented. Sampling time is reduced from hours to seconds in this RSP model.

13.5.2 Sampling on Big Data

Sampling at record level: This is not applicable for large HDFS dataset as it goes through the file record by record. Apache Spark supports RDD. Here sampling function can be categorized as:

- Simple random sampling: This uses sample() function.
- Stratified sampling: This uses functions like:
    - i. sampleByKey()
    - ii. sampleByExact()

Apache Spark makes use of the above functions in batch-wise such that all data items are grouped together and sampling is performed. Besides this, another function called divide and conquer() can also be implemented.

Sampling at block level: Here, a block in place of a record is selected randomly. For the same sample size, it requires less block accesses. More effective result of evaluation can be obtained by using multiple sampling strategies on same dataset.

13.5.3 Sampling-Based Approximation

It uses multi-stage sampling method in order to activate approximation procedure in Apache Spark, Hadoop MapReduce, etc. Early accurate result library (EARL) provides early estimation results. It supports SQL-based aggregation queries. It supports several machine learning models like regression, probabilistic principal component analysis, etc. It enables sampling-based approximation in IoT.

13.6 Big Social Data Analytics

In order to analyze social data, several steps are followed:

1. Data is collected from social media sites using data analytic tools.

2. Correctness of data is checked and security is ensured.
3. Appropriate methods, techniques, and tools are used to model:
   a. Interaction analysis
   b. Conversation analysis
4. Applying methods, techniques from soft-computing (crisp sets, fuzzy sets, rough sets, and random sets).
5. Software realization of findings.
6. Outcomes from conferences, journals, and books.
7. Derivation of benefits in terms of required data, actionable insights, valuable outcomes, and consistent impacts.

13.7 Applications

13.7.1 Manufacturing Production

GMDA as proposed by Hao Zhang *et al.* [11] was universal and did not require skilled experts. Besides this, it had several other advantages like:

1. It could finish the data analytics task automatically with minimum human effort.
2. It provided a language called GMDL which described the tasks that appeared in the process of manufacturing. Without any previous knowledge, non-expert users could easily use GMDL. Hence, it was suitable for small and medium manufacturing companies.

GMDA, a framework, could be better described by three use cases covering all manufacturing processes in the industry:

Use case for inventory forecasting: It would include raw material inventory forecasting. Accurate prediction would lead to a smooth prediction.

Use case for product evaluation: When a new product would be launched, public's demand and comfort should be kept in mind. It extracted rule basing on feedback given by users on various products.

Use case for monitoring tool condition: As manufacturing processes were becoming more complex day by day, real-time monitoring was given more stress. Failures adversely affect reliability, safety, availability, and maintainability.

GMDL provided basic description of task to GMDA. GMDA was based on R language and, hence, not able to handle big datasets. With RHadoop and SparkR, big datasets could be handled based on which work is going on.

13.7.2 Smart Grid

Eklas Hossain *et al.* [4] worked for the application of big data in smart grid. Their findings include:

i. When grid system was combined with power system, information technology, communication, and control system to create an infrastructure, that would support new and emerging technology called smart grid.
ii. Connectivity and exchange of information improved with IoT.
iii. IoT devices generated huge amount of data known as big data.
iv. Big data generated in smart grid required new analysis techniques like machine learning methods for handling and extraction of data.
v. Generation, transmission, and distribution of data in smart grid are under risk. Work is going on detection of cyber security threats and protection mechanism to counter them.

13.7.3 Outbreak of Flu Prediction from Social Site

Balsam Alkouz *et al.* [9] developed a new system named Tweetluenza in order to estimate the spread of influenza in UAE (United Arab Emirates). Their findings include:

i. A multi-lingual and multi-dialectal system which was capable enough to find out a possibility of influenza outbreak from the tweets made on Twitter site.
ii. Used a step-by-step procedure for finding out and apprehending the influenza-related activity throughout the region. It was able to classify tweets of Arabic and English language related to influenza. The classifier was able to remove those words which were not related to influenza.
iii. The processed tweet was assigned proper class label as self-reporting, non-self-reporting, or non-reporting.
iv. All the reporting cases lead to the exact estimation of growth of influenza.
v. Used linear regression on reporting cases in hospital to apprehend the same disease-related cases to hospital in near future.

13.7.4 Sentiment Analysis of Twitter Data

As per Mudassir Khan *et al.* [3] analyzing sentiments was a technique for identifying and classifying tweets from the Twitter site. This provided an overview of public opinion regarding company, product, services, etc. They had used a framework called Hadoop and a classifier called deep learning for sentiments analysis. As the process started, Twitter data was fed as input to Hadoop framework which then worked for feature retrieval. Generally, feature retrieval was carried out in the mapper phase. Then features obtained were fed to the shuffle phase consisting of:

1. Post-map shuffling phase
2. Copying phase
3. Sorting phase

Those unique features that were obtained from this phase were then fed to the reducer phase. This phase used a classifier named deep recurrent neural network (RNN). It classified the characteristics into groups such as:

1. Positive review
2. Negative review

The analysis showed that their work on algorithm based on deep RNN was having maximum efficiency. Future work included more number of features in the said process.

13.8 Electricity Price Forecasting

Kun Wang *et al.* [5] worked upon big data analytics for electricity price forecasting in smart grid as it was cost-effective and older methods might find it difficult to handle massive data related to price in the grid. Hence, a new model for electricity price forecasting was devised consisting of three phases within it. Those phases are listed as:

1. In order to eliminate the feature redundancy, they merged random forest (RF) and Relief-F algorithm so that a hybrid feature selector based on grey correlation analysis could be proposed.
2. The next phase was feature extraction process where an integration of kernel function with principal component analysis (KPCA) was carried out.
3. Use of a differential evolution based support vector machine (SVM) classifier to forecast price classification.

Finally, their results revealed that their proposal had superior performance than other methods.

13.9 Security Situational Analysis for Smart Grid

Smart grid gets benefits from advanced communication and data processing technologies. There is possibility of cyber security threats to smart grid. Currently, smart grid security is based on conventional protection and detection methodologies. These threats normally hamper the normal operation of smart grid. Once threats are detected, it is difficult to repair the damage caused by them. Hence, Jun Wu *et al.* [13] tried to solve the above issue by providing a mechanism capable of creating awareness related to big data security associated with smart grid. In order to carry out analysis task related to security associated with smart grid, they had integrated game theory and reinforcement learning with fuzzy cluster based analytical approach. The final result of their experiment showed less error and high efficiency.

13.10 Future Scope

A person with good analytical skill has ample opportunities ahead in his career. There are several reasons regarding why to opt this field, which are listed below:

1. It is easy to start if one has the ability to learn and do something different.
2. It is a profession in demand.
3. It has multi-domain opportunities like: prescriptive analytics, predictive analytics, descriptive statistics, etc.
4. If one is skilled in this domain, then he/she can draw high salary.
5. It helps in exploring different analytic tools and techniques.
6. Being a big data analyst, one can take decision on behalf of organization based on analysis performed and, hence, have authority.
7. It is not limited to a single language and framework.
8. A big data analyst can easily become a stake holder.
9. There is tremendous growth opportunity due to digitization.
10. One can have a flexible career option.

13.11 Challenges

The main challenge behind rapid increase of volumes of data is not only to deal with it but also to manage rising heterogeneous formats in reference to increasing complex and interconnected data. There is a tremendous need for skilled professionals with analytical skill. A person with high analytical skill can master the world of big data and become an asset for the organization, business, and corporate sector.

13.12 Conclusion

The emerging field of research is the big data analytics. To make the right decision based on the information obtained through continuous inspection of data. It has been adopted by most of the industries. Analytic tools are used by companies/industries to forecast what would occur next. It is assumed that big data is the successor of business intelligence which helps to transform the forecasts into a common language. As this is an emerging technology in today's world, more and more research work are conducted in order to have a better understanding of this.

References

[1] Huangke Chen, Jinming Wen, WitoldPedrycz, Guohua Wu , Big Data Processing Workflows Oriented Real-Time Scheduling Algorithm using Task-Duplication in Geo-Distributed Clouds,IEEE Transactions on Big Data ,Volume: 6 , Issue: 1 , March 1 2020.

[2] Mohammad Sultan Mahmud, Joshua Zhexue Huang, Salman Salloum, Tamer Z. Emara and KuanishbaySadatdiynov, A Survey of Data Partitioning and Sampling Methods to Support Big Data Analysis, IEEEXplore, Volume: 3, Issue: 2 , June 2020.

[3] Mudassir Khan, AadarshMalviya, Big data approach for sentiment analysis of twitter
data using Hadoop framework and deep learning,ic-ETITE, 978-1-7281-4142-8/20©2020 IEEE.

[4] EklasHossain, ImtiazKhan, Fuad Un-Noor, SarderShazaliSikander and Md.SamiulHaqueSunny,Application of Big Data and Machine Learningin Smart Grid, and Associated Security Concerns: A Review, Volume 7, January, 2019, IEEEAccess.

[5] Kun Wang, ChenhanXu, Yan Zhang, Song Guo, Albert Y. Zomaya, Robust Big Data Analytics for Electricity Price Forecasting in the Smart Grid, IEEE Transactions on Big Data Volume: 5 , Issue: 1 , March 1 2019

[6] MiladMakkie, Xiang Li, Shannon Quinn, Binbin Lin, Jieping Ye, Geoffrey Mon, Tianming Liu , A Distributed Computing Platform

for fMRI Big Data Analytics, IEEE Transactions on Big Data ,Volume: 5 , Issue: 2 , June 2019.

[7] MingJian Tang, MamounAlazab,YuxiuLuo, Big Data for Cyber security: Vulnerability Disclosure Trends and Dependencies, IEEE Transactions on Big Data, Volume: 5 , Issue: 3 , Sept. 1, 2019.

[8] Sunil Kumar, Maninder Singh, A novel clustering technique for efficient clustering of big data in HadoopEcosystem, IEEEXplore, Volume: 2 , Issue: 4 , Dec. 2019.

[9] Balsam Alkouz, Zaher Al Aghbari_, and JemalHussienAbawajy, Tweetluenza: Predicting Flu Trends from Twitter Data, Big Data Mining And Analytics,ISSN 2096-0654 05/08 pp273–287, Volume 2, Number 4, December 2019.

[10] Xuedi Qin, YuyuLuo, Nan Tang, and Guoliang Li, DEEPEYE: An Automatic Big Data Visualization Framework, Big Data Mining And Analytics,ISSN 2096-0654 06/06 pp75– 82,Volume 1, Number 1, March 2018.

[11] Hao Zhang , Hongzhi Wang, Jianzhong Li, Hong Gao, A Generic Data Analytics System for Manufacturing Production, Big Data Mining and Analytics,IEEEXplore, Volume: 1 , Issue: 2 , June 2018.

[12] Chunqiang Hu ,Wei Li, Xiuzhen Cheng, Jiguo Yu , Shengling Wang, RongfangBie, A Secure and Verifiable Access Control Scheme for Big Data Storage in Clouds, IEEE Transactions on Big Data,Volume: 4 , Issue: 3 , Sept. 1 2018.

[13] Jun Wu, Kaoru Ota, Mianxiong Dong, Jianhua Li, Hongkai Wang, Big Data Analysis-Based Security Situational Awareness for Smart Grid, Data, Volume: 4 , Issue: 3 , Sept. 1 2018.

[14] YoussraRiahi, Sara Riahi,Big Data and Big Data Analytics: Concepts, Types and Technologies, IJRE, Vol. 5, No. 9, September-October 2018, PP. 524-528.

[15] Dr. S.PraylaShyry, Dhrupad Kumar Das, A Secure And Verifiable Access Control Scheme For Big Data Storage In Clouds, IJPAM, Volume 119 No. 12, 2018, 14147-14153.

[16] Ravi Vatrapu, RaghavaRaoMukkamala, AbidHussain, And Benjamin Flesch, Social Set Analysis: A Set Theoretical Approach to Big Data Analytics, IEEE Access, Volume 4, 2016.

[17] G. Bharadwaja Kumar,An Encyclopedic Overview of Big Data Analytics,IJAER,Volume 10,N0 3,2015,5681-5705.

14

Video Usefulness Detection in Big Surveillance Systems

Hemanta Kumar Bhuyan[1] and Subhendu Kumar Pani[2]

[1]Department of Information Technology, Vignan's Foundation for Science, Technology & Research (Deemed to be University), Guntur, Andhra Pradesh, India, PIN-522213
[2]Principal, Krupajal Computer Academy, India
E-mail: hmb.bhuyan@gmail.com, skpani.india@gmail.com

Abstract

In the past few decades, the safety of human beings and their properties have been issued for survival in real life. The different smart connected communities have taken several steps using prominent components of video surveillance systems for public safety. Sometimes video surveillance system is not working properly to identify the activities/gestures of humans. Thus, the operation and maintenance teams can take the responsibility for significant quantity of time and recognizing the breakdown in a huge video surveillance system. But both useful as well as useless video data consumes data rates through the network and stock up in the cloud.It needs appearance of edge computing for video pre-processing with an edge camera. Thus, it considered the video usefulness (VU) system using special computing system such as edge computing to generate several visual data in huge video surveillance systems. The identification of failure with good bandwidth is considered to find the location of failure and forward to corresponding users. This difficulty is solved by three categories such as (a) calculation of VU values, (b) online failure finding techniques, and (c) good network bandwidth. The above categories lessen overload video data in network based on unused VU values.

Keywords: Video surveillance systems (VSSs), video usefulness (VU), edge computing, failure detection.

14.1 Introduction

Nowadays, public safety is important for human beings when they travel from one location to another for different needs of life in urban area. Due to enlargement of a city area as per population growth,the life protection is vital for city steadiness [1], [2]. The several video surveillance systems (VSSs) [3] are implemented all over the place organized to cover almost cities and their outside neighboring areas, e.g., inner as well as outer roads, traffic, several organization buildings, shopping malls, etc.It is considered for security of public life by controlling illegal activities in city through several high technological devices. For instance, the largest big data is used in surveillance video in extensive city maintaining more than one million video cameras organized [4]. As per extensive VSS, it has considered to take strong steps to remove useless video and to accumulate useful video data using several cameras by maintaining less data stock up usage.

Generally,network camera (NetCam)receives control data and sends image and video data via the Internet. It collects large quantity of video data and forwards to corresponding storage area through cloud for its analysis. The redundancy reduction is accomplished by binary arithmetic coding technique. The measurement of video quality based on its resolution [5], [6] is utilized for processing of video data. Diversity techniques such as video detail description with recovery[7], identification of entity during movement from one place to another using VSS [8], and Synoptic video collected from crowd area based human activities for forensic video surveillance [9]are utilized in the cloud.

The VSS-based storage market product covers storage area network (SAN), network attached storage (NAS), direct attached storage (DAS), and video recorders. The SAN part is supposed to keep huge size of storage for the duration of the estimate period, as SANs are widely used in large enterprises and government organizations for storing video surveillance data and recordings.

The implementation of VSS solutions along with huge enterprises is advanced to generate data based on geo-spatial and user base location. As per public demand as well as security concerns, the high-resolution cameras are used in public places through VSS with all time.Each camera demands enormous storage capacity to stock up surveillance footages for analytical data inspection. Thus, Seagate Technology LLC [10] has used 566 petabytes (PB) of data through video surveillance cameras installed throughout the globe in 2016. It takes the basic challenge to control huge VSS when any failure case occurs. Accordingly, the performance of processing and analyzing huge video data is increased.

Generally, in VSS, few failures of VU may not detect timely. Thus, the number of staffs needs to engage for supervising the identification of failures through quality of service (QoS) [11] and quality of user experience (QoE) [12] from video streams which is more expensive and it extends mean time to detection (MTTD) [13]. Sometimes, the failure video data disturb storage space in cloud storage management.

14.1.1 Challenges of Video Usefulness Detection

The challenges are detected through several questionnaires for solution such as:

(a) What kind of failure status is involved in the huge VSS?
(b) How can one recognize the failure in video data to keep away from deserving the cost of stock up video data?
(c) How is edge computing model utilized to find failures in VSS?

Sometimes, the detection of video failure data is difficult in network, but it can be identified through QoS metrics based on interruption and loss of packet. It doesn't also use error video data in QoS when it is uploaded to the cloud with low quality bandwidth. Useless video data exaggerates over load in the network

The quality of QoS or QoE problems normally occurs in video data. It can take the attractive experience on several video data with QoS metrics,and excluding the whole description of video data is ineffective as per malfunction in camera. The high resolution of videos is well elaborated in [14], and the disturbance of storage in cloud is also explained in [15].Sometimes, the detection occurs false because of human prejudice and making fresh challenges as per demand. In addition, the cloud controls huge quantity of video failure data that makes disturbance of stock up data in cloud.

14.1.2 Video Usefulness Model

The video usefulness (VU) model is developed to identify a failure in VSS cleverly. It focuses the basic objective to create VU model [65] such as:

(1) online fashion is considered to collect video data with package using edge camera and network node;
(2) identifying the failure cleverly on the fly;
(3) forwarding alert information of failure corresponding team and trying to recover it as quick as possible.

For the above design of model, it considered edge computing [16] for huge VSS. It also use edge video analysis for security of people [17], fog computing based on dynamic video stream [18], and also smart urban surveillance [19] in the domain of the Internet of Things (IoT) [20]and Internet of Everything (IoE) [21] in current years.

The edge computing is considered to identify malfunction structure for VU in VSS [65]. This structure completely builds up with computing concept to perceive a failure and lessen the MTTD. This method helps to control several video data rates for storage in cloud. The VU model is established with several parts of VSS system and identifies the process of the corresponding section of VSS. Therefore, VU focuses to meet the goals as follows:

(a) all identified failure in a video flows to lessen the MTTD for a failure;
(b) reduce the trouble of data during transmission and lessen the stock up in effective video data.

The above aim generates an effective manner to develop the quality of video data in VSS and stock up in cloud.

14.2 Background

The background of the above model contains: (1) quality of video services with largest source of video data which is determined and evaluated through QoS and QoE; (2) edge computing for visual video in cloud.

14.2.1 (a) Quality of Video Services (QoS)

The VSSis a huge resource for large visual data storage. The VSSs are contacted with distortion in the video signal to get loss of data during broadcast. It creates unconstructive impact on QoS [11] or QoE [37] of a video data stream. QoS is defined internationally based on several service performances [38].

Generally, QoS parameters are utilized in the area of video service which is suggested by providers. QoS metrics are designed by using several factors such as: error and bit rate, number of task per unit time, interruption during data transfer, complete cycle time, data loss rate, etc.

The above factors are identified edge cameras through compression and distortion. The approaches of QoS determine position for video data and also utilize additional channel to compute the excellence of data service [39]. In addition, QoS approaches cannot determine the entire surveillance video quality. It evaluates only packet loss and the delay.

(b) Quality of Experience (QoE)

It only maintains the quality of video services as per user demand [40].It uses three kinds of approaches such as:

(a) subjective;
(b) objective;
(c) hybrid.

(a) A subjective QoE method [41] is a consistent method to measure the perception of a user. Through this method, it determines the mean opinion score (MOS) [42,43] and appears for media quality. This method is determined by the amount of work as per time. This method does not give assurance on video services evaluation. Therefore, this method does not broadly utilize to determine video services.
(b) Objective QoE approaches [44] exercise several steps to improve the performance of data services through cloud. Thus, the assessment is carried out fast online.
(c) Hybrid approach [45] merges the above two approaches to assess QoE fast and lessens the time and employee resource costs.

Different methods are used to estimate video data quality [41].Video quality and system resources are scheduling of two opponents [46] as per spatial and temporal characteristics. Video quality evaluation models [47], [48] are categorized into three kinds of references such as:

(a) full-reference (FR);
(b) reduced-reference (RR);
(c) No-reference (NR).

(a) FR methods [49] [50] determine the comparison between original images to candidate images based on quality evaluation. This method approaches on base images, where the camera does not stock up whole images. RR methods [51] [52] utilize complete feature data which are taken from base images for determining the deformed image.
(b) NR methods [53] [54] do not use the base images for evaluation. It utilizes data from different pixel domains as well as bit flows of an image/video to assess video quality.

14.2.2 Edge Computing

Edge computing is equivalent to fog computing and cloudlet is a fresh computing model. All appear with the propagation of the IoE in the last few years [55, 56, 57, 58]. In the IoE period, a large number of data will be created by several electronic devices through Internet, which are used in everyday lives, and a large number of applications are used based on the IoE data. The edge computing consists of all kinds of computing over network,Internet,or cloud as per resources.The online video analysis based VSS [59] [60] is much highlighted through use cases for edge computing.

Cloud computing is effectively centralized through large data processing environment [61, 62], but it is not much of a support to VSS through IoE due to the following reasons:

(1) the sufficient computing resources are not centrally available in the cloud to sustain with the volatile increasing computational requirements for large video data through end cameras;
(2) it generates latency between end cameras and cloud during transmission of data;
(3) the confidentiality and protection with data providers at the end;
(4) the power restrictions of the end devices.

The fresh edge computing has the vitalized big data processing for several matters at the end users of the network [63]. Thus, QoS and QoE are two quality aspect model emphases on video quality and its evaluation through video service providers in cloud. The huge number of unnecessary videos is transferred to the cloud to make the irritation in the network bandwidth. But QoS and QoE try to make good quality of VU. It focuses on (a) image/video processing and (b) edge computing.

(a) Image/video processing: This part uses failure detection techniques for images/videos through edge computing model. These computing techniques are used with great real-time performance for video pre-processing to lessen the number of unnecessary videos in bandwidth of network and stock up in cloud.

(b) Edge computing: This is novel computing system for the framework on VU. It permits end cameras to do image/video processing for developing the performance of malfunctiondiscovery.Itused VU analytics through VSS [64].

Thus, it uses a VU framework to develop video quality with analysis. VU aggregates both video/image processing and end user computing for processing video flows and identifies failures of videos through end nodes. VU lessens the data rate and makes maximum video storage in cloud.

14.3 Failure of Video Data in Video Surveillance Systems

In VSS, failure of video data is classified into three groups as per [65]. Those are described as follows.

(a) Failure cases:
A VSS is normally organized with cameras, video servers in the cloud, different end-users, and the network. As per the processing of VSS, an edge camera maintains video data flows through cloudserver for all users who are involved in the system. When failure occurs, it affects several components of VSS during video flows. The VSS uses the video data with the following three qualities:

    (a) High QoS (or QoE);
    (b) Poor QoS;
    (c) Useless content.

For high QoS, video data is used to supervise the scenarios with its analysis. From the above category, high QoS maintains video data analytics that does not affect the two categories. The other two categories do not involve in data analytics. For example, thevideo with useless content through prohibited camera cannot be utilized in video analytics. Poor QoS of video data cannot be also utilized for video analytics.

(b) Component failure:
Several components may occur asfailure in VSS, which is called failure domains through the following components:

    (a) edge cameras;
    (b) cloud servers;

(c) endusers;
(d) network.

In addition, it defines the following failure domains in VSS:

(a) Fedge domain;
(b) Fcloud domain;
(c) Fuser domain.

The above failure domains are briefly described as follows:

(a) Failures through edge cameras are identified as the Fedge domain.
(b) Failures through cloud servers (e.g., meta-data servers)with connected grouping network nodes make Fcloud domain.
(c) Failures occurring through end-user are considered as the Fuser domain.

The above failures can affect the VU.

(c) Video usefulness:
The VU depends on occurrence of failure and its contact on the large number of videos that detail VSS. The VU model identifies the useful video data from any node in VSS through cloud server.

Definition 1: The large number of video flows data is defined as the quantity of practical and malfunction-free portion of whole video flows from cloud data server during time period.

Here, the several mathematical models of video data are explained in VSS [65] as follows.

In VSS, assume *V* to be the number of video data collected from camera during fixed time interval $[1;N_T]$. Based on uploaded video data to a cloud storage server through different nodes, the quantity of video data will increase as per mathematical model. Further, it assumes $q_i$to be the quantity data on videos collected from the *k*th camera through *i*th time. The range is considered as $[1;N_T]$, and the quantity of data on videos is generated by the sum of all q_i as follows:

$$V_k = \sum_{i=1}^{N_T} q_i. \tag{14.1}$$

The d_i video data are broadcasted through *k*th camera and cloud server.Let M_{node} nodes beavailable in the path. Let *U*(*j*) identify the data value of usefulness on vide of rom *j*th node per unit time in VSS. Then,

$$U(j)\epsilon[0;1]. \tag{14.2}$$

From Equation (14.2), if $U(j) = 0$, it is considered as useless for video data from *j*th node and it is not possible to be uploaded in the cloud. If$U(j) = 1$, it uses and uploads video data into cloud.

In this work, the following methods are considered to be developed for managing or handling VU in VSS.

(a) Edge devices have capability to remove video data transmission from cloud as per the VU value.
(b) The VU value will be forwarded to each connecting user for the next decision.

14.4 Approaches of Video Failure Detection

The video failure creates useless content for network bandwidth as well as excess stock up practice in the cloud. Thus, a video failure is identified by testing the data on video in cloud with the help of computing resources. Ituses an edge computing paradigm for each connecting node to describe VU in VSS. The former effort [16] is considered the "edge" for any computational resources through network with the help of the data resources and cloud data centers The failure detection approaches are controlled through the following areas, such as:

(a) establishing practicing resources within the device;
(b) fresh integrated practicing units;
(c) confined practicing resources for cloud with enduser.

The special computing is considered as edge computing units with computational ability for edge nodes. The video streams identify a mal function from all nodes Any camera does not involve with failure detection techniques due to inadequate computing skills. Thus, it considers end cameras with computational unit and is connected to a router. A stack technique is used on a computational unit that establishes the video flows for uploading into the cloud.

Here, three kinds of frame works are considered to identify several failures for VU based on corresponding computing resources above, i.e., (a) edge VU, (b) cloud VU, and(c) user VU frameworks. An edge computational framework is called edge VU. Another two frameworks, such as cloud VU and user VU, are used near resources to detect mal function from cloud and enduser through cloud servers and end user devices, respectively. The frameworks are described below for more information.

(1) Edge VU framework:
Both hardware and software are used for edge VU framework based on an edge camera with few computing resource devices. Most of the hardware platforms have computing power for computational framework, e.g., Raspberry Pi V3 [22], Intel Movidius [23], and NVIDIA Jetson TX2[24]. The messages are transmitted by a VU detector using edge real-time operating system through application interfaces.
(2) Cloud VU and user VU frameworks:
The cloud domain uses the cloud VU framework, where as and the end-user domain uses the user VU framework (individual computers, mobiles, etc.). The above two frameworks use similar structure and identify malfunction data based on closer resources through cloud server or a mobile phone using the VU library.

14.5 Failure Detection and Scheduling

It considers different trivial approaches to identify video failures that vary based on the combined mal function identification techniques. End computational activated techniques are used for video flows.The performance forwards to VU server with endusers.Sometimes the useless video needs to be forwarded by users as in [65].

14.5.1 Failure Detection Approaches in Domains

Different failure detection approaches are developed for each domain. Failure detection is considered to be described in three detection domains such as Fedge domain, Fuser domain, and Fcloud domain, and each domain is also explained with failure description as follows.

14.5.1.1 Failure Detection in Fedge Domain

(1) Solid color screen and freeze frame detection: This detection is used for computational cameras based on its lower cost computing resources. This technique looks for solid color images in video flows and estimates its time. Freeze frame occurs when the last image frame is being repeated. An RGB image is changed to a gray-scale image with each pixel. The gray values estimated through pixels based on degree of the image. The threshold values are used for gray variance [25] for pixels on image with a solid color malfunction. Several comparing threshold values are considered for images. Several variance inter-frames are estimated through reference.

(2) Noise detection: This detection depends on noise failure and casual variation of the informatics data of image. Sometimes Gaussian noise [26] is used for noise detection through normal distribution. Another noise detection like [27] considered as impetuous noise is identified by gloomy pixels in bright regions or reverse in a video. The statistical values are determined between pixels of test video having noise.The statistical values exceed threshold, and then pixels are considered as noise point. When the estimated value exceeds a threshold, it generates noise point. In this case, the corresponding value of the function [28] lessens the suggestion value.

(3) Blurring failure: In blurred image, the sharpness is lesscompared to the original image. It looks likea video through a transparent screen. The edge detection is determined by several blocks of image [29]. To make sharpness (S) image, it needs to calculate the small blocks through fuzzy probabilities.The sharpness value is compared with reference value for failure case.

(4) Angle deviation and scene change detection: The estimated images are used at the border of image to determine an angle deviation failure as entering parameters to speed up healthy function [30]. It used Brute Force Matcher function based on two sets of feature points between the two images. Let two pair matching points be$p_i(x_i,y_i)$ and $p_j(x_j,y_j)$ and used for calculating the Euclidean distance between the two points as $\Delta p_i = |p_i(x_i) - p_j(x_j)|$ and $\Delta p_j = |p_i(y_i) - p_j(y_j)|$, respectively, with calculation$t(p_i, p_j)$. The angle deviation is counted based on exceeding reference values which generate angle deviation failure. Moreover, a failure happens on scene when the normal scene incorrectly changes as per threshold.

(5) Flicker screen detection: Let the consecutive frames f_{i-1}, f_i, and f_{i+1} bepulled out from the video flow to estimate variances of the two inter-frames [25]. A flicker screen malfunction is discovered by two variances with thesuggestion value.

(6) Color cast detection: Any image pull out from video is transferred into the Lab color space[31]. The mean and variance values are helpful to get color cast factor for failure.

(7) Abnormal brightness detection: An estimated RGB image is changed into the hue, saturation, intensity (HSI) color space [32]. The abnormal brightness failure is determined by using number of pixels which exceeds a reference value in the image.Then, the HSI approach helps to make the diffusion and intensity in an image.

(8) Video occlusion detection: In this detection, the reference value of surrounding image to a reference value is collected from the video flows in the first two seconds. The fore ground image [33] is determined by inter-frame difference. After the processing of image, the fresh image is identified as img1Area.Based on the boundary of image growing [34], the larger area of image as maxArea is estimated as per img1Area.

14.5.1.2 Failure Detection in the Fuser Domain

In this part, the malfunction detection approaches are used due to browsing video clips and images by the end-user domain.

(1) Video lag detection: This detection occurs on video lag failure due to network latency between an appeal data packet of end camera and arrival packet. Moreover, the loss of packet is estimated based on mean of response packets and the latency for fast online video.

(2) Mosaic failure detection: This happens based on the poor quality of the broadcast image data in the network. The mosaic-square reference model is considered to identify this failure after Sobel edge detection for estimated image[35].

(3) End-user connection detection: This detection is considered for solid color that identifies the connectivity using the feedback of Transmission Control Protocol or Internet Protocol (TCP/IP) data packets . The cloud domain forwards estimated data packet to each user and maintains several data through VU server.

14.5.1.3 Failure Detection in the Fcloud Domain

Sometimes, humans are facing difficulties to handle huge video data and identify malfunctions online in cloud. If the previous video applies into the server, the video service will be disabled due to failure in a cloud server. It uses three detection approaches to identify locating failures in the cloud such as Fcloud 1, Fcloud 2, and Fcloud 3.

(1) Cloud server connection detection: In this detection, Fcloud 1 is used in two types of network connection: (1)from an edge camera to a cloud server and (2) from an end-user node to a cloud server. The finding system is

Table 14.1 Matching among TP, TN, FP, and FN.

	Positive	Negative
True	√	√
False	×	×

established by an edge domain. For this detection, it uses the feedback from TCP/IP data packets.

(2) Video lag and loss detection: This detection is considered when a video lag failure is found out through each user which is unacceptable in time-sensitive illegal cases. It needs to check the latency through VU library. The latency is determined by the time period between request and arrival packets. The rate of packet loss determines the number of data loss packets for video data. Fcloud 3 is caused the video loss by a severe packet loss.

14.6 Methodological Analysis

Several advantages can be obtained using the computational VU model in a VSS. The detailed description of evaluated data is not required for books. But little explanations on evaluation are given for the sake of understanding as following models:

(a) VU model accurately evaluates the video usefulness;
(b) failure is detected using VU model in a VSS;
(c) lessens the MTTD.

14.6.1 Test of Video Usefulness Model

The performance on VU model is developed through several true and false positive and negative test data items on images and video clips. The accuracy is determined by using binary data based detection techniques [36]. It classifies two kinds of detection outputs such as normal and failure. Several outputs are generated using the VU model detection as follows.

Two important videos such as failure and normal are considered. Two symbols (√,×) are used for matching among failure and normal videos as shown in table 14.1. The four parameters are very important for several estimations. The parameters are true positive (TP), true negative (TN), false positive (FP), false negative (FN).

The dataset with$M$ data items is considered, where$M = (A + B)$,in which$A$ and $B$are considered as the quantity of true failure and normal data items, respectively. Based on TP, FN, TN, and FP, it evaluates accuracy ratios such as efficiency, precision, recall, and specificity on VU model. The efficiency (E)is defined as the mean of recall, specificity, and accuracy on the estimated data as below. Thus,$E$is defined as

$$E = (Recall + specificity + accuracy)/3 \times 100\%.$$

The precision (P)is the measurements of positive values on true and false predictive items. It determines predictive value to evaluate the estimated data nearer to each other. It is defined as

$$P = (TP/(TP + FP)) \times 100\%.$$

The next method, known asrecall,is determined by the fraction of true failure video data items in the dataset with detected failure items and expressed as

$$R = (TP/A) \times 100\%.$$

The specificity (true rejection rate, S) is the fraction of true normal video data items in the dataset being detected as normal items. $S$is defined as

$$S = (TN/B) \times 100\%.$$

In the above methods, 80failure and 50normaldata items with A = (TP + FN) = 80 and B = (TN + FP) = 50 are considered. As per the model, it uses two parameters, i.e., TP and TN, for evaluation. The other description evaluations are not mentioned because experimental resultant information is lengthy. Thus,the chapter is concluded with the above description.

14.7 Conclusion

The VU model is an important model for huge VSS based on vital computing techniques such as edge computing and cloud computing. The VU model focuses on failure detection effectively using huge amount of video data which are shared among edge cameras and end users through the cloud servers to handle huge VSS. Several kinds of failure domains are elaborated through the VU model, and it also estimates the usefulness of video data. Different failure detection approaches are explained to determine the malfunction during video data transmission. Finally,few technical methods elaborated the predicted video items on true or false basis evaluation VU model.

References

[1] T. D. Raty, "Survey on contemporary remote surveillance systems for public safety," IEEE Transactions on Systems Man & Cybernetics Part C, vol. 40, no. 5, pp. 493–515, 2010.

[2] A. Kumbhar, F. Koohifar, I. Guvencÿ, and B. Mueller, "A survey on legacy and emerging technologies for public safety communications,"IEEE Communications Surveys & Tutorials, vol. 19, no. 1, pp.97–124, 2017.

[3] S. W. Smith, "Video surveillance system," U.S. Patent 6 757 008,Jun. 29, 2004.

[4] T. Huang, "Surveillance video: The biggest big data," Computing Now, vol. 7, no. 2, pp. 82–91, 2014.

[5] D. Marpe, H. Schwarz, and T. Wiegand, "Context-based adaptive binary arithmetic coding in the h. 264/avc video compression standard," IEEE Transactions on circuits and systems for video technology,vol. 13, no. 7, pp. 620–636, 2003.

[6] M. Uhrina, J. Frnda, L. Sevcik, and M. Vaculik, "Impact of h.264/avc and h. 265/hevc compression standards on the video quality for 4k resolution," Advances in Electrical & Electronic Engineering,vol. 12, no. 4, pp. 905 – 908, 2014.

[7] N. Dimitrova, H.-J. Zhang, B. Shahraray, I. Sezan, T. Huang, andA. Zakhor, "Applications of video-content analysis and retrieval,"IEEE multimedia, vol. 9, no. 3, pp. 42–55, 2002.

[8] K. A. Joshi and D. G. Thakore, "A survey on moving object detection and tracking in video surveillance system," International Journal of Soft Computing & Engineering, vol. 2, no. 3, pp. 44–48,2012.

[9] B. Yogameena and K. S. Priya, "Synoptic video based human crowd behavior analysis for forensic video surveillance," in International Conference on Advances in Pattern Recognition, pp.1–6, 2015.

[10] M. R. Future, "Video surveillance storage market research report - forecast to 2023," https://www.marketresearchfuture.com/reports/videosurveillance-storage-market-5848.

[11] M. S. Hossain, "Qos-aware service composition for distributed video surveillance," Multimedia tools & applications, vol. 73, no. 1,pp. 169–188, 2014.

[12] M. Li and C.-Y. Lee, "A cost-effective and real-time qoe evaluation method for multimedia streaming services," Telecommunication Systems, vol. 59, no. 3, pp. 317–327, 2015.

[13] T. T. Collipi and D. F. Harvey, "Method and apparatus for analyzing surveillance systems using a total surveillance time metric,"U.S. Patent 7 436 295, Oct. 14, 2008.

[14] T. Kon, N. Uchida, K. Hashimoto, and Y. Shibata, "Evaluation of a seamless surveillance video monitoring system used by high speed network and high-resolution omni-directional cameras,"in International Conference on Network-Based Information Systems.IEEE, 2012, pp. 187–193.

[15] A. Osuna, "Ibm system storage n series and digital video surveillance," Advanced Materials Research, vol. 953-954, pp. 1113–1116,2010.

[16] W. Shi, J. Cao, Q. Zhang, Y. Li, and L. Xu, "Edge computing: Vision and challenges," IEEE Internet of Things Journal, vol. 3, no. 5, pp.637–646, 2016.

[17] Q. Zhang, Z. Yu, W. Shi, and H. Zhong, "Demo abstract: Evaps:Edge video analysis for public safety," in Edge Computing, pp. 121–122, 2016.

[18] N. Chen, Y. Chen, Y. You, H. Ling, P. Liang, and R. Zimmermann, "Dynamic urban surveillance video stream processing using fog computing," in IEEE Second International Conference on Multimedia Big Data, pp. 105–112, 2016.

[19] N. Chen, Y. Chen, S. Song, C.-T. Huang, and X. Ye, "Smart urban surveillance using fog computing," in IEEE/ACM Symposium on Edge Computing, pp. 95–96, 2016.

[20] S. Li, L. Da Xu, and S. Zhao, "The internet of things: a survey," Information Systems Frontiers, vol. 17, no. 2, pp. 243–259, 2015.

[21] A. B. Mutiara, "Internet of things/everythings (iot/e)," in One day Seminar, 2015.

[22] R. Pi, "Compute module development kits now available!" 2017, accessed 14 December 2018.https://www.raspberrypi.org/blog/compute-moduledevelopment-kits-now-available/.

[23] Intel, "Intel Rmovidius TM neural compute stick," 2018, accessed14 December 2018. https://software.intel.com/en-us/movidiusncs.

[24] NVIDIA, "Nvidia jetson systems - the ai solution for autonomous machines," 2017, accessed 14 December2018. https://www.nvidia.com/en-us/autonomousmachines/embedded-systems-dev-kits-modules/.

[25] L. Sendur and I. W. Selesnick, "Bivariate shrinkage with local variance estimation," IEEE signal processing letters, vol. 9, no. 12,pp. 438–441, 2002.

[26] T. Barbu, "Variational image denoising approach with diffusion porous media flow," Abstract & Applied Analysis,2013,(2013-1-30),vol. 2013, no. 1, p. 8, 2013.

[27] R. C. Gonzalez and R. E. Woods, Digital Image Processing.Addison-Wesley, 2010.

[28] M. A. Khanesar, E. Kayacan, M. Teshnehlab, and O. Kaynak,"Analysis of the noise reduction property of type-2 fuzzy logic systems using a novel type-2 membership function," IEEE Transactions on Systems, Man, & Cybernetics, Part B, vol. 41, no. 5, pp.1395–1406, 2011.

[29] J. Gao and N. Liu, "An improved adaptive threshold canny edge detection algorithm," in International Conference on Computer Science & Electronics Engineering, vol. 1, pp. 164–168, 2012.

[30] H. Dong and D. Y. Han, "Research of image matching algorithm based on surf features," in International Conference on Computer Science & Information Processing, pp. 1140–1143, 2012.

[31] M. Kohn, "Method for analyzing images and for correcting the values of video signals," U.S. Patent 6 683 982, Jan. 27, 2004.

[32] N. S. P. Kong and H. Ibrahim, "Color image enhancement using brightness preserving dynamic histogram equalization," IEEE Transactions on Consumer Electronics, vol. 54, no. 4, 2008.

[33] At' .Bayona, J. C. SanMiguel, and J. M. Martt'ınez, "Comparative evaluation of stationary foreground object detection algorithms based on background subtraction techniques," in IEEE International Conference on Advanced Video & Signal Based Surveillance, pp. 25–30, 2009.

[34] J. Tang, "A color image segmentation algorithm based on region growing," in International Conference on Computer Engineering &Technology, vol. 6, pp. 6–634, 2010.

[35] J. N. Sarvaiya, S. Patnaik, and S. Bombaywala, "Image registration by template matching using normalized cross-correlation,"in International Conference on Advances in Computing, Control, & Telecommunication Technologies, pp. 819–822, 2009.

[36] J. H. Saltzer and M. F. Kaashoek, Principles of computer system design: an introduction. Morgan Kaufmann, 2009.

[37] P.-H. Wu, J.-N. Hwang, J.-Y. Pyun, K.-M. Lan, and J.-R. Chen,"Qoe-aware resource allocation for integrated surveillance system over 4g mobile networks," in IEEE International Symposium on Circuits and Systems, pp. 1103–1106, 2012.

[38] C. Recommendation, "I. 350: General aspects of quality of service and network performance in digital network, including isdn,"Marÿco de, 1993.
[39] M. Vranje?s, S. Rimac-Drlje, and K. Grgit'c, "Review of objective video quality metrics and performance comparison using different databases," Signal Processing: Image Communication, vol. 28, no. 1,pp. 1–19, 2013.
[40] K. Piamrat, C. Viho, J.-M. Bonnin, and A. Ksentini, "Qualityof experience measurements for video streaming over wireles snetworks," in International Conference on Information Technology: New Generations, pp. 1184–1189, 2009.
[41] T. Tominaga, T. Hayashi, J. Okamoto, and A. Takahashi, "Performance comparisons of subjective quality assessment methods for mobile video," in International Workshop on Quality of Multimedia Experience, pp. 82–87, 2010.
[42] R. C. Streijl, S. Winkler, and D. S. Hands, "Mean opinion score(mos) revisited: methods and applications, limitations and alternatives,"Multimedia Systems, vol. 22, no. 2, pp. 213–227, 2016.
[43] I.-T. Rec, "800.1: Mean opinion score (mos) terminology," International Telecommunication Union-Telecommunication Standardisation Sector, 2003.
[44] Q. Huynh-Thu and M. Ghanbari, "The accuracy of psnr in predicting video quality for different video scenes and frame rates,"Telecommunication Systems, vol. 49, no. 1, pp. 35–48, 2012.
[45] S. Winkler and P. Mohandas, "The evolution of video quality measurement: From psnr to hybrid metrics," IEEE Transactions on Broadcasting, vol. 54, no. 3, pp. 660–668, 2008.
[46] M. Roitzsch and M. Pohlack, "Video quality and system resources: Scheduling two opponents," Journal of Visual Communication &Image Representation, vol. 19, no. 8, pp. 473–488, 2008.
[47] Z. Wang, G. Wu, H. R. Sheikh, E. P. Simoncelli, E. H. Yang, andA. C. Bovik, "Quality-aware images," IEEE Transactions on Image Processing, vol. 15, no. 6, pp. 1680–1689, 2006.
[48] S. Chikkerur, V. Sundaram, M. Reisslein, and L. J. Karam, "Objectivevideo quality assessment methods: A classification, review, and performance comparison," IEEE transactions on broadcasting, vol. 57, no. 2, pp. 165–182, 2011.
[49] T. N. Pappas, R. J. Safranek, and J. Chen, "Perceptual criteria for image quality evaluation," Handbook of Image & Video Processing, pp. 669–684, 2000.

[50] K. Manasa and S. S. Channappayya, "An optical flow-based full reference video quality assessment algorithm," IEEE Transactionson Image Processing, vol. 25, no. 6, pp. 2480–2492, 2016.

[51] Z. Wang, H. R. Sheikh, and A. C. Bovik, "Objective video quality assessment," Handbook of Video Databases Design & Applications,vol. 17, no. 5, pp. 1041–1078, 2003.

[52] C. G. Bampis, P. Gupta, R. Soundararajan, and A. C. Bovik,"Speed-qa: Spatial efficient entropic differencing for image andvideo quality," IEEE Signal Processing Letters, vol. 24, no. 9, pp.1333–1337, 2017.

[53] H. R. Sheikh, A. C. Bovik, and L. Cormack, "No-reference quality assessment using natural scene statistics: Jpeg2000," IEEE Transaction son Image Processing, vol. 14, no. 11, pp. 1918–1927, 2005.

[54] M. T. Vega, C. Perra, F. De Turck, and A. Liotta, "A review of predictive quality of experience management in video streaming services," IEEE Transactions on Broadcasting, vol. 64, no. 2, pp. 432–445, 2018.

[55] M. Satyanarayanan, "The emergence of edge computing," Computer,vol. 50, no. 1, pp. 30–39, 2017.

[56] F. Bonomi, R. Milito, J. Zhu, and S. Addepalli, "Fog computing and its role in the internet of things," in Proceedings of the first edition of the MCC workshop on Mobile cloud computing, pp. 13–16, 2012.

[57] S. Yi, C. Li, and Q. Li, "A survey of fog computing: concepts,applications and issues," in The Workshop on Mobile Big Data, pp. 37–42, 2015.

[58] K. Gai, M. Qiu, H. Zhao, L. Tao, and Z. Zong, "Dynamic energy aware cloudlet-based mobile cloud computing model for green computing," Journal of Network & Computer Applications, vol. 59,pp. 46–54, 2016.

[59] H. D. Park, O.-G. Min, and Y.-J. Lee, "Scalable architecture foran automated surveillance system using edge computing," The Journal of Super computing, vol. 73, no. 3, pp. 926–939, 2017.

[60] A. Chowdhery, P. Bahl, and T. Zhang, "Bandwidth efficient video surveillance system," Mar 2017, uS Patent 20,170,078,626.

[61] B. Hayes, "Cloud computing," Commun. ACM, vol. 51,no. 7, pp. 9–11, Jul. 2008. [Online]. Available:http://doi.acm.org/10.1145/1364782.1364786

[62] M. Armbrust, A. Fox, R. Griffith, A. D. Joseph, R. Katz, A. Konwinski,G. Lee, D. Patterson, A. Rabkin, and I. Stoica, "A view of cloud computing," International Journal of Computers & Technology,vol. 4, no. 2, pp. 50–58, 2013.

[63] M. Satyanarayanan, P. Simoens, Y. Xiao, P. Pillai, Z. Chen, K. Ha,W. Hu, and B. Amos, "Edge analytics in the internet of things,"IEEE Pervasive Computing, vol. 14, no. 2, pp. 24–31, 2015.

[64] G. Ananthanarayanan, P. Bahl, P. Bodt'ɩk, K. Chintalapudi, M. Philipose,L. Ravindranath, and S. Sinha, "Real-time video analytics:The killer app for edge computing," computer, vol. 50, no. 10, pp.58–67, 2017.

[65] Hui Sun, Weisong Shi, Fellow, IEEE, Xu Liang, Ying Yu, VU: Edge Computing-Enabled Video Usefulness Detection and Its Application in Large-Scale Video Surveillance Systems, IEEE Internet of Things Journal, Volume: 7, Issue: 2 ,Page(s): 800 - 817 Feb. 2020

Index

About the Editors

Subhendu Kumar Pani received his Ph.D. from Utkal University ,Odisha, India in the year 2013. He is working as Professor and Principal at krupajal Computer Academy,Odisha,India. He has more than 17 years of teaching and research Experience His research interests include Data mining, Big Data Analysis, web data analytics, Fuzzy Decision Making and Computational Intelligence. He is the recipient of 5 researcher awards. In addition to research, he has guided two PhD students and 31 M. Tech students. He has published 51 International Journal papers (25 Scopus index). His professional activities include roles as Book Series Editor(CRC Press, Apple Academic Press,Wiley-Scrivener),Associate Editor, Editorial board member and/or reviewer of various International Journals. He is Associate with no. of conference societies. He has more than 150 international publications, 5 authored books, 15 edited and upcoming books; 20 book chapters into his account. He is a fellow in SSARSC and life member in IE, ISTE, ISCA,OBA.OMS, SMIACSIT, SMUACEE, CSI.

Somanath Tripathy received his Ph.D. in Computer Science and Engineering from the Indian Institute of Technology, Guwahati, in 2007. Currently, he is an Associate Professor in the Computer Science and Engineering Department of the Indian Institute of Technology, Patna. He joined the faculty in December 2008. His research interest includes lightweight cryptography, Security issues in resource restrained devices, Blockchain and Machine learning Security. He has published more than 80 research papers in different journals and conferences of repute. He has been PI of several projects related to security. Dr. Tripathy is currently serving as an associate editor of Sadhana-Academy Proceedings in Engineering Science of the Indian Academy of Sciences and editor of IETE Technical review.

George Jandieri finished secondary school in Tbilisi with a Gold medal. In 1966 -1971 he continued his studding at the Tbilisi State University where he graduated with honors, Theoretical Physics. In 1978, he defended his Candidate thesis in the special field of Radio-physics at the Tbilisi State University. In 1979, G. Jandieri was awarded a candidate degree in physical-mathematical sciences by The Highest Certifying Commission of the USSR. He defended his doctoral thesis in Radio Physics-Plasma physics in 1989 at the Kharkov State University, Ukraine. The authority of this scientific council was determined by an active participation of the famous physicist-theoretician, a Nobel Prize winner Academician, Lev Landau. George Jandieri was awarded the honorary title of a Senior Scientific Worker (1990), a Doctor's degree of Physics-Mathematic Sciences (1991), and the honorary title of Professor (1991) by The Highest Certifying Commission of the USSR. Prof. George Jandieri is the Academician of the Georgian Ecology of Science; member of the Institute of Electrical and Electronics Engineers (IEEE) since 1998. In 1999-2006 he was an expert of the International Association for the Promotion of Cooperation with Scientists from the Independent States of the Former Soviet Union (INTAS) in Europe; since 2012 he is Euro-Expert. In 2003-2005, 2012-2013 he was foreign expert for the Italian Scientific Projects and in 2005 was certified as The Highest Category International Expert. In 1998, George Jandieri was elected as the Deputy Director General of "Who's Who in the World" in Europe (office is located in London, University of Cambridge). He was awarded diplomas and a medal from this Organization as well as international certificates.

Sumit Kundu (SM'07) received the B.E. (Hons.) degree in electronics and communication engineering from National Institute of Technology, Durgapur, India, in 1991, and the M.Tech. degree in telecommunication systems engineering and the Ph.D. degree in wireless communication engineering from IIT Kharagpur, Kharagpur, India, respectively.,Since 1995, he has been a faculty in the Department of ECE, NIT Durgapur, where he is currently a Full time Professor. His research interests include wireless ad hoc and sensor networks, cognitive radio networks, cooperative communication, energy harvesting, and physical layer security in wireless networks. As of today, he has published more than 150 research papers in various journals and conferences., Dr. Kundu is a Reviewer of several IEEE and Elsevier journals.

Talal Ashraf Butt received the Ph.D. degree in Internet of Things from Loughborough University, U.K. He was with the 5G Innovation Centre (5GIC), the U.K. Government's funded initiative at the University of Surrey to develop 5G technologies. At 5GIC, he gained experience of working in a testbed team that owned the state-of-the-art mobile testbed and successfully developed and demonstrated novel research ideas. He is currently an Assistant Professor with the American University in the Emirates. He is a Reviewer of IEEE journals. He is passionate about next generation networks and protocols.